QA 76.9 H85 B88 2009 c.2

Studies of Work and the Workplace in HCI

Concepts and Techniques

Synthesis Lectures on Human-Centered Informatics

Editor
John M. Carroll, Edward M. Frymoyer Professor of Information Sciences and Technology, Penn State University

Common Ground in Electronically Mediated Communication
Andrew Monk
University of York, U.K.

Studies of Work and the Workplace in HCI
Graham Button, Sheffield Hallam University, and
Wes Sharrock, The University of Manchester

Semiotic Engineering Methods for Scientific Research in HCI
Clarisse Sieckenius de Souza and Carla Faria Leitão
Pontifical Catholic University of Rio de Janeiro (PUC-Rio)

Information Cultivation for Personal Information Management
Steve Whittaker
University of Sheffield, U.K.

Copyright © 2009 by Morgan & Claypool

All rights reserved. No part of this publication may be reproduced, stored in a retrieval system, or transmitted in any form or by any means—electronic, mechanical, photocopy, recording, or any other except for brief quotations in printed reviews, without the prior permission of the publisher.

Studies of Work and the Workplace in HCI: Concepts and Techniques
Graham Button and Wes Sharrock
www.morganclaypool.com

ISBN: 9781598299878 paperback

ISBN: 9781598299885 ebook

DOI: 10.2200/S00177ED1V01Y200903HCI003

A Publication in the Morgan & Claypool Publishers series

SYNTHESIS LECTURES ON HUMAN-CENTERED INFORMATICS #2

Lecture #2

Series Editor: John M. Carroll, Penn State University

Series ISSN

ISSN 1946-7680 print

ISSN 1946-7699 electronic

Studies of Work and the Workplace in HCI

Concepts and Techniques

Graham Button
Sheffield Hallam University

Wes Sharrock
University of Manchester

SYNTHESIS LECTURES ON HUMAN-CENTERED INFORMATICS #2

ABSTRACT

This book has two purposes. First, to introduce the study of work and the workplace as a method for informing the design of computer systems to be used at work. We primarily focus on the predominant way in which the organization of work has been approached within the field of human–computer interaction (HCI), which is from the perspective of *ethnomethodology*. We locate studies of work in HCI within its intellectual antecedents, and describe paradigmatic examples and case studies. Second, we hope to provide those who are intending to conduct the type of fieldwork that studies of work and the workplace draw off with suggestions as to how they can go about their own work of developing observations about the settings they encounter. These suggestions take the form of a set of *maxims* that we have found useful while conducting the studies we have been involved in. We draw from our own fieldwork notes in order to illustrate these maxims. In addition we also offer some homilies about how to make observations; again, these are ones we have found useful in our own work.

KEYWORDS

work, workplace, organizations, ethnomethodology, ethnography, situated action, interactionism, design, computer supported cooperative work, human–computer interaction

Contents

Introduction

This book is an extended version of a chapter entitled "Studies of Work in HCI" first published in the work of Carroll (2003). The extension relates to three of its parts. The first is a more detailed discussion of the provenance of studies of work and the workplace in human–computer interaction (HCI) and extends the "scientific origins" section of the original version. In the original version, there were three sections: "ethnography," "ethnomethodology and conversation analysis," and "situated action." This extended version retains those sections, but considerably elaborates upon them. We have done this because those whose work forms the backbone of our discussion of studies of work and the workplace in HCI are thoroughly immersed in these origins, yet most considerations of studies of work do not articulate them in any depth. Many of the people whose work we refer to take these origins utterly for granted, and, therefore, if people coming to studies of work for the first time are to properly appreciate them, a more thorough understanding of their intellectual origins is appropriate. For those who may just be interested in studies of work as part of a rich methodological array in HCI, however, this chapter may be more than they require. But for those who may be embarking upon studies of their own, this chapter is intended as a foundation upon which a deeper understanding of what is involved can be built.

The second and third extensions to the original chapter involve two new sections, which are specifically oriented to those who are intending to develop studies of work themselves. We are often asked by students who are embarking on fieldwork or ethnography for the first time, "How do you do it?" We take it that this question has nothing to do with their reading and understanding of studies of work or of the literature behind them, but the practical question of how they take that understanding and make it turn in the real world they may observe around them as they develop their own studies. We have therefore introduced two new sections into the original consideration, which articulate some analytic *maxims* that, in the course of our own investigations, we have used to open up the settings and the activities and interactions we have witnessed, and some *homilies* on how to collect data from those settings and of those activities and interactions.

CHAPTER 1
Motivation

Grudin (1990) argued that human–computer interaction (HCI) had passed through a number of stages in its development and was, at that time, moving from the fourth stage, which he characterized as "a dialogue with the user," to a fifth stage, which is focused not so much around the individual but around the *work setting*. In the intervening years since he made this point, there has indeed been an increasing emphasis in HCI on research and also the development of systems that support the interactions and collaborations between people in their workplace. This has been particularly so in the field of computer supported collaborative work (CSCW). Part of this research has involved turning to the disciplines that have been traditionally associated with research into the work setting, most notably sociology; consequently, HCI researchers have increasingly looked to sociology for their understanding of work and the workplace (cf. Sommerville, Rodden, Sawyer, & Bentley, 1992). In addition, sociologists have increasingly become directly involved in HCI and CSCW research.

Sociology, however, is far from a unified field and it is possible to see that a variety of sociological theories and methods have been courted and used as resources in HCI (Shapiro, 1994). Dourish and Button (1998) have argued, however, that it is one particular type of sociology that has come to predominate within HCI concerns with the workplace. It uses an ethnographic/fieldwork approach rather than drawing off a particular social theory. It is empirical in as much as it is concerned with the *analysis* of work and the workplace. It also owes much to the methodology of ethnomethodology and conversation analysis. It is the manner in which studying work under these auspices has been developed within HCI that is referred to in this book. Studying work, in this manner, has become in the last 10 years a general method in HCI design and development, especially in the area of CSCW.

There are a number of precipitating developments that have led to its adoption. The publication of *Plans and Situated Action: The problem of human–machine communication* (Suchman, 1987) was a particularly defining moment for a significant minority in HCI, especially those concerned with the then fledgling interest in CSCW. Suchman's attack on the cognitive science understanding of human doings resonated with those within HCI who were attempting to move away from the predominate focus on the *user* and were attempting to place the computer within a social context of *use*. A cognitivist approach may, on the face of it, have enabled HCI to address the individual's use of the computer; however, it is a limited resource for the consideration of social relationships despite recent but unconvincing forays into distributed cognition theory. Suchman's critique of cognitivism, however, went much deeper than pointing out its limited scope. She mounted an attack on iconic figures in cognitive science, such as Herbert Simon, for their failure to understand the implication of the fact that human action is situated in the social and cultural world for the explanation of that action. In particular, she undermined the idea that human action can be accounted for in terms of mental predicates, such as plans, which have become an important concept in the development of artificial intelligence. Suchman's rebuttal of such an explanatory framework made reference to the fact that all action takes place within a swarm of socio-cultural contingencies that cannot be covered in full and in advance by a plan, an argument that owes much to ethnomethodology.

A second precipitating development to the adoption of studies of work was the very development of CSCW itself. CSCW was emphasizing the design of interactive systems. In part, this was recognition of the fact that many undertakings and people's working environment involve collaboration, and the concertion of actions, something which the traditional HCI cognitive approach was not well equipped to handle. Suchman provided CSCW researchers with a theoretical underpinning for their position. However, Suchman did not just provide an anti-cognitivist argument, she also furnished *studies of use*. These studies, however, were not done within the confines of the traditional experimental situations characteristic of much HCI research at the time, but involved studying use in the actual workplace in which systems were being used. At this time, Suchman's work drew heavily from the fields of conversation analysis and ethnomethodology within sociology. Suchman was thus furnishing those within HCI who were dissatisfied with its individualistic and cognitivist approach with a powerful critique of cognitivism and the beginnings of a new empirical method of study.

A third precipitating moment in the development of studies of work was the Scandinavian Participatory Design movement (cf. Greenbaum and Kyng, 1991). Members of this movement had, for a number of years, argued that the requirements for technology should be developed directly around the work situation of the technology's users, and by the direct involvement of the people who would use the system in the design process. The Participatory Design movement had developed methods and perspectives on interactive systems design from this position that emphasized issues such as the flexibility involved in work activities; the idea that work is an accomplished rather than a mechanical matter, and that the workers' voice should be heard with respect to workplace management and development. Again, Suchman's emphasis on *work practice*, an idea derived from ethnomethodology, resonated with their practical attempts to develop interactive systems for the workplace.

The study of work as a method in HCI has proliferated in the past 10 years. Not all studies are conducted under the influence of Suchman's work, nor under the auspices of ethnomethodology, although most would claim to be ethnographic. This book will not cover the varied ways in which work has been studied in HCI but will concentrate on the attempts to develop a method for HCI that involves studying work for the purposes of rigorous interactive systems design, drawing off the methods of ethnomethodology and conversation analysis as developed in sociology. It does so because the resulting analytic orientation to studies of work is by far the most systematic and developed of the studies of work undertaken within HCI, and certainly the most prolific.

* * * *

CHAPTER 2

Overview: A Paradigmatic Case

One out of what we will later see are a number of ways in which studies of work have been applied in human–computer interaction, has been to analyze the impact of a particular system on the ways in which that work is organized in the work setting into which the system was introduced. This has often led to a critique of a system for its dysfunctional consequences on the organization of the work and the setting. Studies of work have thus been used to analyze the organizational principles or methods behind a domain of work, to analyze the impacts of a system upon these methods, and to critique its design when it conflicts with these methods. These methods and critiques then become available for the design of subsequent technologies.

One family of technologies in the workplace that has attracted a considerable interest in HCI (see Suchman, 1994, and the subsequent replies and commentaries in vol. 2, nos. 3–4, 1994, of *Computer Supported Cooperative Work*) is concerned with the coordination of work. Often, it has been found that these coordination technologies clash with the methods through which a domain of work is itself coordinated. The designer, rather than studying this "natural" coordination, has often produced an idealized model of coordination which, when applied to the work setting, overconstrains the work, with the result that it becomes more difficult for people to accomplish their work assignments using the technology. Coordination technologies that have been developed to better support this type of work have thus been criticized for having the opposite consequences, making it more difficult for people to coordinate their work. A study by Bowers, Button, and Sharrock (1995) is a case in point.

Bowers et al. studied the introduction of a production management system into the print rooms of a number of production printers in the United Kingdom. They studied the methods through which production was organized before the introduction of the system and then, once the system had been introduced, analyzed how it played into these methods. A principle and an intended outcome is that the natural coordination methods used by print operators, administrative staff, and management was the smooth flow of work across the print room floor in such a manner that machines and operators were fully occupied and that the jobs they were working on were produced on time. However, after the introduction of the new system, this smooth flow of work was severely disrupted; during the first month, many sites were significantly behind on their workload and had failed to meet the contracted deadlines of many of their customers. In order to work around the problems the system had caused, the management in many of the sites disabled the system, although it had cost them many tens of thousands of pounds to install.

The problems encountered could not be passed off as "teething problems." Rather, the reason for the problem was that the designers had inappropriately modeled the workflow that the system enforced. On the face of it, it might seem that the relationships between the various stages that make up printing are simple and straightforward. A job is first processed by the administrative staff who, among other things, give it a job number. It then goes to the origination department, which sets it and does the art work. From there, it is moved either to plate making if it is a wet ink job or directly to printing if it is to be electronically printed. If the job goes to plate making, it then goes to printing; all printed jobs are then passed on to the finishing department, where they are cut and bound. The last stage is dispatch.

All of the print sites studied by Bowers et al. had a number of processes that controlled the relationship between the various stages and which are routinely used in this type of business. The production management system was designed to control the workflow of the job through the various stages in accordance with these processes. However, Bowers et al. found in their study, and before the introduction of the system, that in order for the processes to be used in actual situations, it is necessary for all involved to exercise their judgment as to how the processes can be fitted to the occasion. In this respect, a number of practices not covered by the processes have developed across sites in order to achieve the result the processes have been designed to achieve, but which are not on this occasion due to contingencies. These ad hoc practices are familiar methods that have developed in print shops to make the formal process work on the ground, so to speak. The use of these methods is an important factor in achieving the result desired by the formal process, which is the smooth flow of work across the print shop floor that maximizes the operator's time and machine capacity.

In this respect, Bowers et al. described a number of methods that machine operators, administrative staff, and management used to fit the processes to the circumstances. For example, sometimes operators would "jump the gun" and start a printing job when that job had not been assigned a job number. However, according to the formal processes, operators were supposed to wait before they commenced printing, for a job number allocation by the administrative staff. Sometimes, this process could lead to delay and interrupt a smooth flow of work, when, for example, a job could go straight to printing without the need to pass through the origination department. This could happen when an operator had completed their current job but had no next job to work on because the administrative staff had not yet completed allocating it with a job number. To overcome this delay problem, operators would have to see what jobs lay on the administrators' desks and which ones were awaiting processing. As these types of print shops involve repeat runs, it is easier for the operators to find jobs that they already knew the details of, jobs they could set up and start running. They were thereby jumping the gun because they were running the job before the administrative staff had completed their work. Although they were breaking the process, they did so for the jobs that were, practically speaking, unaffected by the process. They exercised their skilled judgment in the circumstances and were able to achieve the result the process was designed in part to achieve: the smooth flow of jobs across the print shop floor in such a manner that operators and machines were working to full capacity.

The system that the company introduced to support production management and work coordination had a formal model of the workflow; it enforced the relationship between the tasks that made up the work of the print room. Thus, the system enforced the formal processes used in coordinating the relationship between the various parts of print production. For example, under the regime of the system, it was now not possible for an operator to start work on a job unless it had a job number. The reason for this was that the operator had to log onto the system by entering his/her identifying password and by entering the job number. Unless both were entered, the system would not recognize the operation. Operators could have, of course, worked on the job, but the records would show them to be idle during that time. Previously, they would just have entered the job into their paper logbook as soon as the job number became available in order to put the record straight. The system overly constrained the operators, which led to production problems. All the staff involved were prevented from exercising their judgment in situations that confronted them and this resulted in the problems that Bowers et al. witnessed upon its introduction. Due to these problems, the system was at best only partially implemented.

Not only does this study of work analyze the methods through which a domain of work is organized by those party to it from the inside, as Bowers et al. put it, it also makes clear that modeling a workflow using a formal process as a resource only partially grasps the work a system is meant to automate or support. This example may be an extreme one, but the general point can be taken: studying work reveals a domain of work practices and methods that are crucial for the efficient running of an organization, which were left unremarked upon by the methods the designers of the system had followed and may go unremarked upon by other methods. Bowers et al. were treating work as a methodical matter and were trying to uncover the methods through which people order their work on the ground, so to speak. By surfacing these methods, they were able not only to critique a bad design and a bad design methodology that overformalized the work, but also to make these methods available for the purposes of design.

* * * *

CHAPTER 3
Scientific Foundations

Studies of work in human–computer interaction draw off a number of interrelated analytic and methodological developments in sociology and anthropology. These are, primarily, ethnography, ethnomethodology and conversation analysis, and the idea of situated action.

3.1 ETHNOGRAPHY

3.1.1 The Tradition of British Social Anthropology

The beginning of modern anthropological ethnography is often counted as beginning with the stranding of the Polish anthropologist Bronislaw Malinowski in Melanesian New Guinea as a result of the outbreak of the First World War ("British social anthropology" is "British" by virtue of its association with the major British universities—London, Oxford, and Cambridge—rather than because of the national origins of its important practitioners). Malinowski was thus forced into prolonged and relatively close contact with those native to the island, and the 1922 publication of his *Argonauts of the Western Pacific* (Malinowski, 1967) is often treated as the first contribution to ethnography. Ethnography, as Malinowski understood it, was a matter of "detailed, first-hand, long-term, participant observation fieldwork, written up as a monograph about a particular people." (Malinowski quoted in Atkinson et al., 2007, p. 60)

Malinowski's study was, of course, not the beginning of anthropological study but was considered a significant advancement; it involved observation of the people under study as they actually went about their lives and engaged in the activities—especially those of religion and ritual—that the anthropologists were interested in. The prior practice had been to rely in "informants," with the study involving a transaction between the anthropologist and one or two senior figures from the tribe who would be interrogated about their practices. Thus, what the members of the tribe actually did was being reconstructed from what the informants said they did, the informants report being accepted as an accurate description of the doings of the tribe. The contact between the anthropologist and the tribal seniors was one that took place within the context of a colonial system. It was not expected that colonialists would have close personal contacts with "natives" (or even "savages") except at the level at which official contact between the colonial system and its indigenous subordinates would take place: colonial official to tribal leader.

Hence, Malinowski being stranded (for some 2 years) by the outbreak of war thus enforced the situation of contact with the tribal peoples and the opportunity to do little else except pay some attention to what the natives were doing around him.

His thinking certainly featured a theme that will recur which is that understanding people's activities and ways involves attempting to understand things from "their point of view" and in "their own terms" and to grasp the native's point of view and his relation to life, and to realize *his* vision of *his* world. We note that the masculine pronouns here indeed refer to males with the tradition of social anthropological study, thereafter taking for granted that the public and institutional life of the tribe was in the hands of males and focusing very much on that with the result that the rise of contemporary feminism was able to raise serious objections to that tradition's basic conceptions.

However, for Malinowski, the understanding of the native's point of view was intended as materials for the formation of a scientific scheme for the understanding of societies. 'The point of view's' role is to show what is seen from close up and through the eyes of a participant; the things people did would appear differently as compared to someone whose only contact with them was from a social and geographic distance and who would often know these events only through reading. A theoretical scheme would be needed to interpret the documentation, one that made Malinowski a progenitor of an outlook that is now known, and usually reviled, as "functionalism." This type of approach, although not derived specifically from Malinowski, was, in the 1930s, adopted as the basis for the close study, in a program known as "the Hawthorne experiments" (Gillespie, 1993), of shop floor industrial workers that became a basis for much subsequent management thinking, and which was to be an early form of the often fruitful idea of "bringing anthropology home" to study parts of Western societies in the same ways they studied the tribal groups in other societies.

Another strand that will run through the discussion here is that of the 'instrumental conception of things'. There is ingrained in our own culture a strong inclination to measure the things that people do against a certain standard of practicality. Is what people do useful in a practical sort of way? If it is not, then perhaps they ought to give up doing this useless thing.

The anthropology of Malonowski laid down important foundations for ethnography that have relevance for systems design. First, understanding culture is important for understanding the actions and interactions of people. If we are to place computer systems in the workplace, then understanding the ways of that workplace is an important resource in the design process. Studies, such as the one conducted by Bowers et al., discussed above, of the introduction of systems into the workplace, document the difficulties that people have had in using a system because it does not fit into the social organization of that workplace.

Also at times, there has been, what can seem to those not involved in design, arrogant arguments in some parts of the design world that in developing new workplace systems, they are reinventing the work and the workplace. Indeed, it has been presumptuously argued that we do not need to know about how things are today in the workplace because designers are changing that world. A classic example of this is the argument around "the paperless office." New electronic technologies, it was argued, would do away with the need for paper. Electronic files could be easily circulated and read on the screen, paper would be a technology of the past, and people's working patterns and relationships would change. Therefore, it is not necessary to know about how people are currently working because one of the points relating to the introduction of new technology is to actually change these ways of working. However, it is abundantly clear that the paperless office has not materialized. Anyone working in an office environment knows this. Studies confirm what everyone knows that there are indeed more printed pages of paper now than before the advent of new office technologies, and some office activities are still best done through the medium of paper. Perhaps if the champions of the paperless office had reflected more deeply on how office workers engage in their work activities and interactions, then the gap between their predictions and what has resulted might have been less (cf. Sellen and Harper, 2002).

However, it could be argued that there are other examples that show how the organization of work has been fundamentally changed through the introduction of new technology, to such an extent that an understanding of its previous organization would be irrelevant for the development of that technology of change. For instance, what organizations now have the typing pools, which were once a prominent part of, for example, print works where there would be constant demand for retyping submitted materials? Studying the organization of the typing pool would serve little purpose since it has turned out that they disappeared. The organizational structure of many institutions now no longer includes such departments. But in changing the organizational structure does the actual social organization of actions, and interactions at work change? Organizational structures come and go, and departments transform, but in doing so, has the nature of the organization of work changed? For example, many law firms once had typing pools; these days, that would be a rarity or possibly non-existent. But in doing away with this means of producing documents, has the organization of the work of lawyers changed? Studies of solicitors show how their work is organized in the organization of files. Thus, files are often organized in terms of steps to be taken in processing a case and relevant documents that represent a step that has been taken. Viewing a certain file can lead to the next activities to be done, thus organizing their files in terms of case processing steps and checks provides for what the solicitor will actually do regarding the case. The organization of files facilitates doing certain actions in an orderly methodical way. (cf. Travers 1947)

The introduction of new office technologies may change the personnel who create the files; the location where they are created; the media of their creation, storage, and conveyance. Solicitors themselves, rather than typists, may use computers to build up the files, which may be electronic rather than paper, and transfer them through networks rather than by hand (although even a casual inspection of some solicitors' offices may actually call into question this degree of change). However, in these changes, has the nature of the logic of the work, its fundamental organization, changed? No, cases still progress in a stepwise fashion, moving onto the next step once the previous step has been completed, that is, the steps are transcribed in advance within a sequential order of activities. What may change is the scenery, but the play of work is of the same methodical order, which is being made visible in the ethnography of the workplace. Knowing that order, designers are in a better position to understand the requirements of a system and in a better situation not to commit the blunders of the past that have led to systems cutting across that order with sometimes disastrous consequences.

Another lesson that designers can learn from the ethnography of British anthropology concerns who it is that designers engage in developing an understanding of what is required of a system. It has been the case that systems designers have engaged in dialogues with the customer for a system in developing this understanding. The customer has typically been the purchaser of the system. Thus, the purchaser, rather like the senior tribal member that anthropology used to understand a culture, has provided an understanding of the work that the system will automate or support. However, hierarchies can be a barrier to a proper understanding of a culture. Within organizations, information seals prevent the circulation of knowledge about the organization moving up and down the hierarchies. Those lower in the hierarchy can have a vested interest in not having those higher up in the hierarchy properly understand what they are doing. It gives them, for example, organizational space to arrange their working lives to their advantage. To those higher up the hierarchy, knowledge is often power and holding on to that knowledge is a way of holding on to that power. Moreover, people in different positions in a hierarchy often cannot understand what the other person does in sufficient detail to capture the high level, let alone the detailed matters of each other's job. In a large corporation, does the Chief Technology Officer (CTO), who is often the purchaser of new technology, sufficiently understand, for example, the work of the dispatch department to safely specify the requirements of a new dispatch system? The head of the dispatch department might, but then he/she might not want the CTO to actually know about some of the practices of their department; thus, even if the CTO and the head of dispatch were informants for the designers, they may still not gain sufficient understanding of the work they are supporting or automating with new technology.

One of the lessons that "classical anthropology" teaches is that in order to understand the culture, do not rely on an informant, go out and be in that culture, follow through what people actually do, not what they or others say they do. If ethnography is to become a paradigm in the design of computer systems, then understanding just this matter is important. Fieldwork is not just talking to a particular group of people in an organization; it is not just interviewing them. Yes it involves that, but that is not the end of the matter. It involves moving beyond what they say in order to see what is done. The lesson of classical anthropology that "studies of work and the workplace in HCI" draw off is that if we are to design systems for the workplace, then designers need to get into that workplace in order to see that culture, not just hear about it. Of course, this involves matters of practicality—how practical is it for designers to do this—and considerations of how to support the systems designers have developed. One example is the system developed by Sommerfield and Martin (this can be accessed through the University of Lancaster website), which provides the designer with a summary of issues developed from studying certain domains of work that they can then use in developing systems to be used within that domain.

3.1.2 The Chicago School of Sociology

In the 1920s and 1930s, there emerged within sociology a particular line of thought that was associated with the University of Chicago that was known as the "interactionist tradition." One of the key figures in what became known as the Chicago School of Sociology was Robert Ezra Park, a former journalist who sought to introduce the "muckraking" traditions of U.S. journalism into the academy, helping it to contribute to an understanding of the process of urbanization, which was then very rapidly transforming the city of Chicago. Park worked in close collaboration and with complementary interests to another colleague, Ernest Burgess. The basic idea was that the city was a "melting pot" created by diverse types of groups, yet its development reduced the diversity between them. Therefore, this process could be captured in its formation in Chicago, which was drawing in all types of racial and ethnic groups; however, this would involve identifying the diversity as early as the stage of city formation, if possible (Park, Burgess, Mackenzie, & Janowitz, 1984). The "muckraking" journalists had gone forth to investigate and find out things for themselves. Park urged his associates and students to follow the same policy to acquaint themselves with the different ways of life that were then lived in, and that were making up, the motley of diversified communities comprising the city. The impetus was to study the lives of the "lower classes," including those who were recent immigrants (as manifested in the monumental study *The Polish Peasant in Europe and America*, Thomas and Znaniecki, 1920; and in studies of blacks who had migrated in from the South, and of the Jewish ghetto), but, and more so, those lives might be taken as symptoms of the disruptive effect of the impact of mobility and social transition on previously settled ways of life.

There was an objective of understanding the developing dynamics of the city as a whole, but this was to be partly effected by the core contribution of ethnographic studies. Although the city as a whole might have a type of general organization, this organization was made up of a diversity of localities. While one effort was to understand the way in which the localities are related to one another and constituted the general pattern of urban formation, it was, of course, necessary to understand those localities themselves. It was here that the ethnographic studies came in.

During the process of formation, the city was a motley of transitional areas, ones that—as their name indicated—were localities with social differences and some of the diversity was temporarily maintained. The "natural" development of these areas would create a range from the most privileged to the most marginal. Thus, attention was often focused on those who were living at the margins of respectability, who were homeless, who were involved in prostitution or near prostitution, and who were involved in crime or gang violence. A particular thrust of the Chicago tradition was to challenge the notion that change and disruption resulted in social disorganization, that those who "misbehaved" did so because they were unable to conduct orderly social lives. The Chicago tradition rebutted this idea, showing that close acquaintance with the lives of those who were supposedly "disorganized" showed that they were not in fact living chaotic lives, but orderly ones—ones that did not conform to middle class standards of life, perhaps, but that were no less intelligently arranged and managed than middle class ones. Furthermore, the life was collectively organized, lived according to conventions that were characteristically shared by the community of individuals involved in these activities.

The ethnographic investigation of the diversity of local communities was conceived as an extension of a basic feature of human life, and one which reflects its essentially social character, that of taking the point of view of the other. The idea is that the individual does not have a "purely" individual point of view, entirely distinctive from that of others, since part of maturation as an individual means incorporating views that other people also have, including the view that they take toward oneself. Human beings, unlike other animals, can "see ourselves as others see us" and can indeed see how others see things. Learning to do activities (as when children are playing at, for example, delivering the mail) involves learning to see how other people see things, for playing "mailman" involves being able to see things in somewhat the fashion that real mailmen do. Being able to interact with others involves being able to anticipate how they will act toward oneself, and this involves being able to grasp how the other person will understand one's own action toward them—since it is on the basis of this that they will act back toward one. Being able to see oneself as another sees one is spoken of as the capacity to take oneself as "an object." Thus, how other people see things and how they see oneself is incorporated into one's very development as an individual personality, and this is the basis for one's capacity to act toward others. However, one's social experience is necessarily limited, especially by the social divisions that make certain groups and ways of life virtually unknown to one, except perhaps through the grossest stereotyping. Thus, ethnography is part of coming to understand how the inhabitants of those communities see things and, of course, with "relativizing" effect, how they see you, too.

The Chicago School of Sociology demonstrated that it was possible to develop interesting insights into and understandings of one's native culture through the deployment of the fieldwork techniques of anthropology. This was certainly a decisive movement in the development of sociology in-as-much as it provided an alternative means of apprehending social organization to that of theorizing or applying statistical analytic techniques upon data gathered about social movements. It was also a decisive moment for ethnography itself in-as-much as it became apparent that anthropological investigation could "come home" so to speak and be undertaken within one's own culture. However, neither of these moments may have a relevance for the undertaking of ethnography for the purposes of developing design methodology. Those designers who are interested in developing ethnography as a paradigm for design methodology take it for granted that it can be used within "domestic" situations, and it may be only of historical interest how that move from "other" to "domestic" culture took place.

Nevertheless, in the actual studies that emanated from the Chicago School, there are a number of matters that designers could heed. In the hands of the Chicago School, fieldwork became a way of progressively developing a familiarization with the concrete details of a diversity of social environments. There are two matters here. The first is the idea of concretization. Like the classical anthropological studies, the studies of the Chicago School dealt with what was actually happening as opposed to what was said to be happening. The importance of this was that all too often what was said to be happening was actually said by social theorists, and the myths that the Chicago School evaporated concerning the social organization of the marginalized were often myths that had in part been spun by other forms of social enquiry to that of ethnography.

Design can be the victim of social theory. We have previously attempted to argue this with regard to Habermas' theory of action, which was proposed as appropriate for the analysis and design of groupware systems (Sharrock and Button, 1997). Basically, our argument there was that Habermas' theory was flawed in itself and therefore basing any categorization system on it for the purposes of analyzing and designing systems was problematic. If the theory is questionable any design based on it is also questionable.

Therefore, concrete investigation is not theorized versions of what people do. The second matter is that of diversity. Many social theories of organizations and the workplace seem to treat them as a homogenized totality. Thus, at a simple level, all management roles and activities are lumped together as "management," as are non-management roles and activities. As a homogenized category, it can easily be assumed that the requirements of individuals within that category are similar. Yet, it is often the case that there can be much diversity between the individuals or organizational parts that are collected together under the category. Within organizations, there can be a diversity of social environments each with potentially different ways of viewing and understanding the organization. To not appreciate this may lead to only a partial or even a particular understanding of the organization being developed. The lesson that we can learn from the Chicago School in doing ethnography as part of the design process is to look for diversity and to understand that diversity may generate more complex requirements for a system; if there is diversity, we should not be satisfied that we have a sufficient understanding of the organization and the workplace from the point of view of "management," for example, because within management there may be a number of points of view.

The very issue of trying to understand social organization from the point of view of those within it is emphasized by the Chicago School. However, it has been particularly deepened in a further contribution to the intellectual development of ethnography by interactionism.

3.1.3 Interactionism

The aim of placing the point of view of those who inhabit the social settings at the center of attention was propelled by a contemporary of Park and Burgess at Chicago, Thomas (1966), who proposed the maxim that if "men define situations as real, they are real in their consequences." The expression "definition of the situation" consequently became a key phrase. First, individuals react to situations as they see them. How they see situations does not depend on the situation alone, but on the assumptions and responses that the individuals bring with them into the situation. Benjamin Lee Whorf, a famous anthropological linguist who had earlier been an insurance investigator, was tasked with finding out why there were explosions in oil storage depots. He found that a major cause was employees taking their cigarette breaks near where the used oil drums were stored. The smokers regarded those drums as empty, but they were in fact "full" of explosive vapor. The workers defined the drums as safe to smoke around and this resulted in high rate of explosions. If you change their definition and make them recognize that the drums were empty only of oil, not of petroleum vapor, then you could alter their reactions to those stored drums.

The situation, the stack of stored drums, does not, of itself, provide the workers with a basis for action; it is their conception of the drums—as either "empty and therefore safe" or "full of vapor and therefore dangerous"—which determines whether the workers will be cautious and refrain from smoking in their vicinity or not. How individuals define their situation varies; different individuals may interpret the same situation very differently and will therefore respond to it in correspondingly different ways (which is one reason why law like generalities about how individuals react to certain types situations will be hard to come by).

Another sociologist who is relevant here is Herbert Blumer (1969), who emphasized the following point very strongly: the organization of conduct involves "interpretation"—the participant in the situation has to draw upon whatever understandings have been accumulated to figure out what the situation is, as a basis for grasping what next action would be appropriate. Situations are not interpreted by the individual alone, but the "definition of the situation" is a matter of joint action, something done collectively and collaboratively among some number of individuals. Obviously, defining a situation is less problematic if it is pretty much a familiar one, one that can be readily defined by being identified as "the same as yesterday"; therefore, situations of change and disruption may be expected to be more problematically defined and there was, therefore, an attendant interest in phenomena of "collective behavior" when groups of people confronted by previously unknown situations attempt to define them—panics, the spread of rumor, the formation of mobs, etc., could be understood in this way. Thus, if people define a situation as real, this will shape how they react to it. It does not matter if, from some scientific or sociological theorist's point of view, individuals wrongly define their situation; it is on the basis of the way they do define it that people will act, and, therefore, that definition of their situation will be real in its consequences—if a mob wrongly defines somebody as, say a pedophile, and are violent toward that person, then that definition will have real consequences for that individual, regardless of whether that person really is pedophile.

Exploring the definition of the situation and the means whereby individuals singly and, above all, jointly "interpret" their situation to arrive at a definition of it became the way in which the understanding of the participant's point of view was construed, and the central aim of fieldwork was to capture the series of actions making up *the process* through which individuals *in relation to each other* defined and redefined their situation. Thus, the business of defining a situation is not a once-and-for-all affair, but an ongoing matter, for an initial definition of the situation may or may not be sustained by further developments in that situation—what may initially seem like a friendly encounter may, as it goes along, turn out to be rather different than it first seemed. Thus, situations unfold themselves, and they are monitored as they do so to detect incongruities with the initial definition, ones that may either manifest that the initial definition was inappropriate or involve a change in the nature of the situation.

The emphasis on joint determination of the definition of the situation and of the sharing of definitions naturally drew attention to interaction between individuals as a communicative medium and to the role of language in both defining the situation and in conveying their respective understandings in arriving at the definition. Blumer phrased it this way: "language is not a neutral medium." He thereby drew attention to the way in which the language of a group reflects the life of the group and incorporates understandings that have been built up by the group and that reflect its interests, concerns, and prior experience; a grasp of these will feed into both the way that the individual reacts to situations and the way in which the actions of each individual affect the actions of others. Thus, although people are nominally the speakers of the same language, it does not follow that they thereby easily understand each other, for what seems to be the same language may actually vary a great deal between social groups. People who are English speakers may nonetheless have access to different "argots," terminologies that are relatively private to their subgrouping, understood within that, but not outside of its boundaries. The emphasis on the way in which people interact with one another on the basis of their interpretation of the situation, what others are doing and the like, became known as interactionism.

Part of the HCI community has developed an interest in the development of systems to support collaborative work (computer supported collaborative work). Interactionism is pointing to a relevant issue here—the collaborative development of understanding within a community—which becomes relevant for how that community works and how the actions and interactions of those within it are organized. Collaborative definitions of the situation need to be attended to. Interactionism becomes even more important in understanding the development of interest in studies of work and the workplace in HCI; it was particularly developed within a genre of interactionist studies of work and occupations, and it is here that interactionism can, more obviously, attract the attention of the systems designer.

3.1.4 Work and Occupations

Everett Hughes (1971), another interactionist, turned attention toward occupations. The reason for this is that one of the main bases for collective organization is within the world of work: there is both the occupational grouping and the work organization. Work plays a central part in the organization of the life of the society and of the individual.

Society involves, as the theorist Max Weber (1964) emphasized, a distribution of "esteem," and one's place in this arrangement very much affects how one thinks and feels about oneself. If others think that you are a worthless, no-consequence kind of person, then this is how you might come to think of yourself—after all, part of what you are is what you are capable of in relation to other people. Can you strike up a friendly relation with them, or can you only approach them in humiliating ways? At the same time, being looked down upon by others does not automatically result in accepting the same lowly estimation of yourself as they make, for people can, either individually or collectively, develop defences against such estimations and can distance themselves (in some ways at least) from the general rank order of the society.

Mentioning the way people think about you and the way in which you think of yourself must not be allowed to mislead, and it should be recognized that this way of stating the point is nonetheless meant to highlight the ways people think of you invokes the ways in which they treat you. For a clear example, two researchers in this tradition, Anselm Strauss and Barney Glaser, (Glaser and Strauss, 1964), were doing fieldwork in U.S. hospitals, where they found out that there were times when the medical staff needed to "ration" their efforts (Becker et al., 1977). They were very busy and had not only too many sick people to deal with but also too many people at risk of death, and they were expected to save all their lives. They had to make a decision as to where they put their effort—who should they try to save. Their decision making gave Strauss the idea for "social loss." They would decide who to try to save on the basis of who they thought would represent the greatest loss to others: who would miss this person if they died, how much the person is missed, etc. Thus, a young mother with young children would get high priority in the distribution of life-saving efforts, whereas someone that "nobody would miss" such as a "derelict" could even just be left on a trolley in the corridor to die, receiving no treatment at all. This orientation is not just and necessarily something imposed upon one by others—think of the "women and children first" policy that used to and probably extensively still does exist, the willingness of people to risk losing their lives by giving others priority where there is an opportunity of escape.

In the modern world, one's occupation is a major, but of course not the only, consideration in fixing one's place in the rankings of esteem and in one's own sense of self-worth, and it was not surprising that one main strand of the inquiries that Hughes and his students set in train was therefore into the world of occupations. A natural object of attention was the increasingly prominent development of occupations that called themselves "professions." Those occupations that had achieved professional status or that were competing for it had rationales as to why they deserved the exceptional respect that a profession has relative to other occupations—the kind of work they do, the responsibility that doing that work demands, the need for a high standard of educational preparation could all be appealed to. Professional occupations were very much independent of other occupations and authorities: the professional practitioner was trusted to get on with his/her work and left alone to do so.

In the professions' own rationales, the privileged positions were the means by which the profession ensured the dependable delivery of a valued service. Hughes pondered, however, whether the privileged position of the profession was perhaps the end, rather than the means? Rather than assuming that professions were distinctive in the ways that they proposed, in the ways that earned them their privileged position in the hierarchy of work, Hughes thought it sensible to investigate whether professional work was really distinctive from that of other occupations that did not have this prestigious standing. He and his students argued that the distinctiveness of professions was often, at the least, something of an exaggeration. Take the case of legal work, for example. This seems like work that requires a highly developed technical competence and one that can only be acquired through extensive legal training. But in that case, why is so much of the actual work of a legal office done not by the professional lawyers but by people in clerical jobs, work that very often involves considerable legal technicalities? The rationale for professional status thus came to be viewed as more akin to self-promotional propaganda than something to be taken at face value, and the professions were seen to be distinctive not by virtue of having such distinctive features as in being successful over other occupations in the competition for higher status.

The "distrust" of the appearances of professionalism reflected a deeper strand in the approach being taken, which was one that involved a formal analytical mentality. The conventional distinction between the professional occupation and others, in terms of highly developed skill, educational preparation, etc., was to be viewed as a sociologically superficial one. Taking that distinction at face value would lead to an expectation that a whole range of differences would exist between a professional and a non-professional occupation that might not be found. There must be commonalities between occupations obtained because these are the constant accompaniments of work, rather than because of the specific type of work that is involved. The similarities and differences between different types of work need to be conceived in more specifically sociological terms for sociological purposes, rather than being marked in terms that have often, and rather trustingly, been taken over from the occupations themselves.

The study of work and occupations, and of organizations, therefore, emphasized the need for a corrective study of those in the lower and most demeaning of occupations, those who, for example, are assigned the task of doing society's "dirty work," work that those in higher status positions would regard as demeaning to do themselves (e.g., garbage collection, keeping disorderly populations such as vagrants and the mentally ill under control) but nonetheless must be done. The point of view of those in lowly positions in the occupational hierarchy was no less worthwhile and effectively equivalent, in sociological terms, to that of their presumed superiors, the inclination to cut across the established order of social evaluations with sociological equivalences. Rather than thinking that the dividing lines of society are drawn to distinguish (and separate) populations with different types of characteristics, the Symbolic Interactionist (SI) approach assumed that people on either side of a social division are apt to be much the same types of people and that the differences between them and those "on the other side of the tracks" are apt to result from, rather than be the cause of, those divisions.

Thus, the necessity to perform demeaning work is treated as something that potentially "contaminates" the individual's self and that potentially makes them think of themselves as "a lesser kind" of person than others; certainly, this is how they are looked on by those who look down on them, as comparatively "inadequate" individuals incapable of properly reasoned thought, to effectively manage their own affairs, and to lead an "organized" kind of life. Let us remember a basic supposition of the school, that is, how you think of yourself is often a function of how you are regarded by others. The study of those in lowly positions assumes and is meant to bring out the following facts: that they are not in fact any "different kind of person" than their presumed superiors; that they are no less capable of reasoned, intelligent, and practically effective conduct than anyone else; that they do not necessarily reciprocate the view their superiors take of them; that they can be skeptical and cynical about those "superiors"; that they contrive ways of "distancing" themselves from the "contaminating" effects of their lowly position and the demeaning things they are called upon to do. They cannot perhaps fully insulate themselves from being at the bottom of the social heap, but they do not have to concur in the evaluation of them that otherwise might follow from this fact.

This may be all well and good for those interested in sociology, but for those interested in design, how does this interactionist interest potentially play out? Well, first of all, it can be seen in these types of interactionist studies that if we view work as just a chain of tasks, for example, we are missing the point as it is seen by those doing the work. Viewing work as a set of tasks is a way in which HCI has often gone about the business of developing requirements. But work is not just definable tasks; work is what people are as well as what they do. Understanding this is important if we are to develop systems that change what they do, for it may mean that people are being forced to change who they are. Resistance to the introduction of new technology in the workplace can thus stem from that technology resulting in unacceptable changes in self-perception or the perception that others have of somebody. Work is not then just some mechanically embodied activity that takes place within a defined geographical local, it is a way in which people define themselves and each other. To understand this, it is necessary to look at work from the point of view of those who do it.

For example, a study by Becker et al. (1977) of the workings of a hospital in the United States forcefully undermined the picture of hospital life traditionally held. That picture bought into the professional view of doctors as having high professional status and thus holding the reins of power within hospitals, especially regarding determinations concerning patient care. However, Becker et al. describe that far from being powerless because they were in a more "lowly" position than doctors, nurses, in fact, exercised considerable power and were as equal drivers of patient care as were doctors. This is because the nursing staff were intimately involved with the patients and had detailed knowledge of their situations. They also held information regarding the patient. Orders might flow down, but orders needed information and information did not necessarily flow up. Thus, patient care and decision making in the hospital involved a complex interaction between groups of people, an understanding of which was plain to those within the work setting but, due to commonly held views on professions, were not plain to those outside of the situation. If systems are to support work, it is important to see that work is not task-based; it is organizationally embedded, something done by people whose investment in doing the work conditions the effectiveness and productivity of the work, and has cultural meaning. Work and peoples' occupations are then complex social matters, and the success of automating or supporting work can in part turn on understanding that.

3.1.5 Redefining Organizations

Work often takes place inside of organizations, and ethnography has been keen to take account of the workings of organizations. One of the major developments here is associated with Goffman's (1961) treatment of "total institutions," a way of reclassifying organizations that initially infringes the kind of moral differentiation that is thought to set them apart. Thus, we might think of monasteries and prisons as being morally very different types of organization, one made up of the best type of people, the other filled with the worst. Thus, we would be normally inclined to start from this kind of difference and to attribute all types of features of the organizations differentially, supposing that, in the case of mental hospitals, it is the inmates' mental condition that explains any strange behavior, and assuming comparably in other cases that it is the moral character of the people in the organization that explains why they act as they do. We would see any facts about a monastery *in the light* of the fact that the monks are somewhat saintly individuals, and the facts about the prison *in the light* of the delinquent nature of its inmates. This may, however, distort the understanding of the ways in which behavior is actually organized. It presumes that patterns of behavior stem from the moral character of the individuals but this may not be so.

The idea for total institutions arose from Goffman's field study in a large state mental hospital and its consideration of the problem of "mental illness." It is easy to see that the idea that people in a mental hospital are "mentally ill" would provide the obvious and ready-to-hand explanation for their patterns of behavior.

However, Goffman argues that many features of life on ships, prisons, monasteries, and mental hospitals are the same regardless of the (presumed) difference in the moral character of those inhabiting the organization. They are "the same" in respect of what we can call "their form." Of course, there are lots of differences between the patterns in the institutions; life in a monastery is not exactly like life in a mental hospital. In both, however, there are lots of ways in which the place is organized to deprive people of their individual identity; monks and the mentally ill are provided with standard issue garb that is the same in appearance. For different purposes and with different techniques admittedly, the two institutions both involve the attempt to deprive individuals of the sense of the distinctiveness and importance of their individual sense of identity. Thus, they do so may not follow from the purposes they serve, so much as it does from the fact that you are dealing with a population that is resident together throughout the daily round. The ways in which organizations "homogenize" their inmates in this context may not be because of their religious, rehabilitative, or therapeutic purpose, but for administrative reasons, in response to the necessities of handling a population that is coresident together throughout a 24-hour period and with very restricted opportunities for individual privacy.

Goffman's approach exemplifies very well the idea of the formal/analytic treatment (and his account of total institutions was continuous with the systematic working out of an array of very different topics around the idea that people's interactions are communicative and that the possibilities developed in those interactions are therefore subject to their capacity to control what information they control about themselves). Thus, his most famous book, *The Presentation of Self in Everyday Life* (Goffman, 1959), draws a parallel between theatrical and social settings more generally. The back stage/front stage division, between what the audience can see and what is being done to keep the play going onstage that is concealed from them, is treated as a feature that can be applied to any social organization, identifying the division between what those involved are doing and what those they deal with can be allowed access to. The analogy also applies to individuals, in the sense that individuals try to keep from others things that are "discrediting" about them and need therefore to be able to control, as they often cannot always effectively do, what they communicate about themselves (hence, "the presentation of self"). Another book, *Stigma* (Goffman, 1963), is concerned with those individuals afflicted by "discrediting" characteristics (such as manifest facial disfigurements or visible signs of membership in a despised minority grouping) and who have to find ways of dealing with the fact that they cannot conceal these from others (as somebody with a former conviction might) and cannot therefore control what these communicate to others. The total institutions case is continuous with this for a feature of these for their "inmates" is that there is little, if any, opportunity for privacy, and the "front region–back region" distinction common to most establishments does not apply to them. Thus, in the "asylum," inmates cannot keep their past histories secret since these are known through the organization's records on them and are made public in the course of therapeutic occasions.

Thus, the idea of the front region–back region distinction is one that cuts across all the divisions that we ordinarily make between different types of organizations, and the adoption of the idea gives us a means of looking at otherwise very different activities as confronting and dealing with the same general problem, that is, of managing, preserving, and coping with breaches of the separation of front stage from back stage. Thus, Goffman's own method involves using many different examples from many different types of situation—as total institutions feature ships, concentration camps, military academies, monasteries, etc.—but this is to show that problems of social interaction and organization have the same "form" although expressed in very different "materials."

The use of the concept of total institutions, by giving us a view of the way in which the situation of the inmate in the mental hospital is formally similar to that of people who are not considered mentally disturbed in any way, enables us to see the situation of the mentally ill and the character of psychiatric treatment in a very different light.

Rather than make the obvious and inviting assumption that the conduct of the inmates of the asylum is to be understood as a product of their inability to behave normally, Goffman preferred to view those inmates as if they were psychologically normal. Rather than viewing what they did as "inappropriate" ways of behavior, Goffman could then ask whether what they did was entirely appropriate to the circumstances in which they found themselves. Could anyone behave any differently to the way the inmates did if they were placed into similar circumstances, regardless of their prior life history?

The point of view of the psychiatric staff is one which is directed toward focusing on the conduct of the individual in isolation from its context; what the individual does is to be understood as an expression of the individual's inner psychological state, and attention is therefore directed away from the context to which the individual may be responding (and to which the therapeutic staff may in any case be insensitive to). The circumstances with which the individual is confronted is imposed upon him/her and is one over which he/she has little control: they have been deprived of many of the resources that they use in their lives in the outside society to organize their affairs. Thus, what they do may constitute perfectly rational responses to their need for privacy, self-esteem, sex, string, or sealing wax in an environment where these cannot be obtained in any "normal" ways and can involve behaving in ways, cut off from an awareness of their purposes, that may well seem odd and inappropriate to the circumstances their "guardians" perceive them as responding to.

Furthermore, a new view of therapeutic practice itself is generated. Those carrying out the practice think of what they are doing as being for the inmate's own good. They are attempting to repair psychological damage; they may well be aware that the inmate does not see things from their point of view and that the inmate does not welcome their treatment in this way. The inmate may have a very strong sense of resentment about their position, feeling that he/she may feel has been let down by those nearest and dearest to them who have collaborated with the authorities to commit them to the hospital and that any "unreasonable" behavior in their relations with others was not their fault and was down to the others rather than themselves. Being in the asylum is doing them no good to the extent that they are cut off from their normal lives, are subjected to various types of humiliating treatment in the hospital, and have acquired an identification as mentally ill, which may make life subsequent to release much more difficult in various ways.

However, their protestations against this will be programmatically treated as symptomatic of their psychological difficulties. The therapeutic staffs have no concrete access to the actual historical events that led to the inmate's arrival in the hospital. However, they do have a policy of disbelieving the claims inmates make and of regarding denials of mental illness, of the legitimacy of confinement, and so forth as further signs of the illness that has brought them here. The application of the therapeutic policy is experienced by the inmates as coercive rather than therapeutic. It is a feature of the therapeutic line that admitting that one is mentally ill, no matter how strenuously one has denied that one is and on whatever "reasonable" grounds, is a sign of progress. It is only when sufficient psychological progress has supposedly been achieved that the inmate will be released from the hospital. Therefore, the inmates realize that it is only when they start to talk in the ways the staff require them to and start saying that they recognize that they are mentally ill that they are going to get any closer to getting out. Thus, inmates may admit that they are mentally ill because their insistence on what they believe is true and just may protract their stay in this undesirable place. Many of the medical and care techniques are not used by the staff in either a therapeutic or a caring way; they are used in a disciplinary fashion to control and punish troublesome inmates.

All social communities have ways of controlling unruly members of the community. It may well be thought that the inmates of a mental hospital will be troublesome and that there will be a need to restrain and punish those who are troublesome. However, the point that is being made here is about the way in which supposedly medical and therapeutic treatments are used not for actual medical reasons, but for punitive reasons only. Locking people up in isolation, putting them in restraints, drugging them, removing their teeth, and bathing them in ways that are physically painful are all done ostensibly as part of the therapy, yet in reality they occur because the staff have had enough of the trouble the inmate is causing them. Thus, there was a type of background view to Goffman's studies that psychiatric treatment may often be just an alternative to the identification of people as criminals or other types of outcast and that psychiatric categories and classifications are ostensibly medical in character. Yet, what they actually classify as "sick" people do not have actual disease or physical affliction but are in one way or another people who do not stick to the rules of the society. Putting people away in prisons or in mental hospitals are alternative ways of achieving the same objective, which is to "exclude" those who have been causing trouble in a community.

The main point we are stressing from Goffman's work is that organizational context is important for understanding the behaviour of those encompassed by it. However, we should not necessarily think that we understand that context because, the organisation is a familiar one. As Goffman shows, it is the underlying form or structure of the organization that matters, not its presumed status. Thus, to understand activities such as work, it is necessary to understand the organizational context and that it is a matter for investigation.

3.1.6 Subcultures

In the manner of the tradition of the original Chicago School, attention was also paid to those who were low in the occupational hierarchy, not least, since they might not fit into this or at least not fit into the conventional occupational hierarchy. There were studies of "social outcasts," such as the mental patients discussed above and other "deviants," those who lead unconventional lives, engage in criminal activity, or in other ways do not fit the "straight" life. Here, the assumption of "sociological equivalence" across social dividing lines resulted in the often misunderstood ideas of what was called "labeling theory" (Schur, 1971). Several different sociologists of this school devised different versions of much the same idea. Traditionally, the idea had been that the way to understand crime was to understand what it was that made those who went in for it different from the rest of us. Thus, there is regular recourse among biologists and the newspapers that read their press releases to the idea that there is a "gene for criminality" or a "gene for homosexuality," etc. The idea is that criminals and homosexuals are different types of people from the rest of us, to the extent of being physically different. The so-called labeling theory opposed the idea that people who are in socially demarcated and exclusive groupings are so because they are basically different from each other—the differences that there are between them are consequences of being classified in different social categories and groupings, rather than the other way around.

Thus, one version of the labeling theory is provided by Edwin Lemert in terms of a distinction between "primary" and "secondary" deviations. Primary deviations are things that people do in infraction of established social rules such as taking drugs when these are outlawed, drinking alcohol when this is subject to legal prohibition, talking too loudly and violently in public places, standing too close to people when talking to them, etc. Secondary deviation is the kind of rule infringing behavior that arises from and responds to primary deviation; this involves people in, for example, organizing to maintain and support their primary deviations, as when people who smoke marijuana have to develop further illegalities in the smuggling and supplying of dope. People break social rules for many and varied reasons. Business people break many laws as nothing more than another economically motivated decision continuous with the legal ones that they make: the risk of prosecution, fine, etc., being merely another business risk along with all the others. Thus, people on both sides of social divides are basically the same type of people; however, their behavior varies because of the different situations that their respective positions put them in, and this results in studies that tend to "deflate" the image of the respectable (reminding us, e.g., of the extent to which, and without damaging consequence, the middle, especially business, classes engage in "white-collar crime") and to "inflate" that of those who are "outsiders" to the mainstream of middle class life.

The point here is that we often assume that we know about people by knowing their "label." Thus, we know what judges do because we know them as judges; we know what ambulance dispatchers, air traffic controllers, stockbrokers (just to list some occupations for whom systems have been developed that have devastatingly failed) because we know them *as* ambulance dispatchers, air traffic controllers, and stockbrokers. The study of subcultures suggests the contrary; it emphases that we do not necessarily know what people or groups of people are like without studying them.

3.1.7 Conclusion

The studies of work and the workplace that have been undertaken for the purposes of addressing the design of computer systems do not overly articulate many of the issues we have considered here. However, they are all ethnographic in nature. Those not involved in ethnography may be aware of its anthropological roots, but as we have been attempting to show, what ethnography is in the human sciences is much more rich and textured. Those who are conducting studies of work and the workplace are variously drawing off this rich tradition, and it is important to understand this if the provenance of these studies and thus their particular interests are to be fully understood.

We can summarize the important steps that have been taken over the years in the development of ethnography in the social sciences as follows:

- The British tradition of anthropology—studies from within the culture;
- The Chicago School—bringing anthropology home;
- Interactionism—what people do is based on how they interpret situations and actions of others;
- E. C. Hughes—the study of work;
- Looking at work from the inside—from the point of view of those who do it;
- It is more than just work—the distinction between what is technically necessary and what is socially invested in;
- The hidden culture of the workplace—cannot be seen from a distance;
- Studying the disposed and the marginal—to get away from the idea that you already know what they are like;
- Individual characteristics are endowed by organizational needs for control—what someone may be doing may be a response to the situation they are put in;
- Discovering local competencies—people actually know what they are doing.

3.2 ETHNOMETHODOLOGY

This book is particularly concerned with studies of work and the workplace in HCI predominantly from the standpoint of ethnomethodology. Ethnomethodology has some affinities with Symbolic Interaction (SI) and also originates in deep reflection on the theoretical and methodological difficulties of sociology, but it draws much more upon the European phenomenological tradition (as initially mediated by Schutz, 1967) than upon the American pragmatist one that had been the inspiration for much of SI's impetus. At the same time, it had a concern akin to that of the positivists for rigor in the treatment and analysis of data, a conviction that a lot of what goes on in sociological investigation could be tightened up right away, rather than at some remote future date.

Simplifying, we could say that the motivating theme was much the same as that which bothered many mainstream sociological theorists and methodologists, which was the tendency for sociological theory and empirical research persistently to come apart. The founding, and still central, text for this approach is *Studies in Ethnomethodology* published by Garfinkel (1967), and a look at the papers in that collection will readily reveal that the main concern is with the methodological problems that were troubling American sociology during the 1950s and 1960s. Many sociologists came to think that the central problem of sociology is that of "objectivity," and that, perhaps in two senses. In one sense, the question is, "Is society something that exists independently of individuals' beliefs about it?" or "Does it consist in individuals' beliefs about it?" In other words, is society an "objective" phenomenon existing in its own right, or is it a "subjective" one existing only in the beliefs of individuals? In The *Social Construction of Reality*, Peter Berger and Thomas Luckmann (1966) argued that sociology had been divided between two rival traditions, one holding that, "society as is an objective reality" and the other that and "society as is a subjective reality." Berger and Luckmann claimed that these traditions were complementary rather than opposed and that society was *both* "an objective" *and* "a subjective reality." There is another sense of the objective–subjective contrast, though, one in which objectivity pertains to that of method, to the idea that sociology's research procedures need to be such as to ensure intersubjectivity, to ensure that there is uniformity in the ways in which individual researchers respond to instances of phenomena, such that they would describe or classify the same instance in the same way. The aspiration to "objectivity" here involves the elimination of subjectivity, which, in this connection refers to the personal and idiosyncratic. The aim of method is to eliminate any tendency toward individual response from the investigative process, to provide impersonal procedures, thereby thoroughly standardizing the collection and interpretation of data

Ethnomethodology is often thought of as being basically on the subjective side of these contrasts; that it is even an extreme advocate of the idea that society exists only in the minds of individuals and is constituted by their beliefs about it. Similarly, it is thought to encourage an approach to research that puts the researcher's own self-consciousness at the center of the investigative process. There might be some truth in such suppositions, but there is not very much, and continuing to think of things in this manner only results in a fairly refined reconstitution of sociological issues being thought of in a crude and simplistic way. Rather, ethnomethodology's key concern might be with the problem of objectification (rather than that of objectivity), with capturing materials that genuinely reproduce the features of the phenomena one intends to study. That is, how do you ensure that the *materials* (or data) that you collect really do display properties of that which you intend to study? How does one find "the society" in the materials that the researcher collects through participation in the society itself? One of ethnomethodology's solutions was to use mechanical recorders such as the audio and video recorder to capture behavior, because, for example, from the point of view of analyzing people's talk a good audio recording captures in considerable detail what was actually said.

It is certainly true that there is little real conflict between sociological theories and empirical research, and that it is only in the weakest ways that one can say that research plays the part of, for example, testing theories as the majority of the discipline suppose that it ought to do. However, Garfinkel noted another significant and persistent difficulty in sociological research: that there tended to be a gap between the phenomenon purportedly studied and the phenomenon actually studied. In other words, the problem was that of "objectification," of collecting data that captured and exhibited the socially organized properties of the intended phenomenon.

The topic of suicide provides a nice example. The study *Suicide* was certainly intended by its author, Emile Durkheim (1951), the founder of French sociology and a founding figure for contemporary sociology more generally, as a canonical example of what a sociological study might be like and was extensively taken to be so because of its use of, albeit in only a basic way, what came to be known as multivariate statistics, one of the staple tools of much sociological data analysis. Durkheim's study was supposedly of suicide, but, of course, Durkheim did not observe any suicides. He presented extensive statistics displaying the varying suicide rates found in different countries and among different types of social groups, but, given that he did not observe any suicides, how did he obtain his data? From the statistics assembled by coroners and other types of legal functionaries, who had not, of course, observed any suicides either.

First then, the notion of an equivalence class is important to the would-be quantitative tradition that Durkheim did much to found, but the idea of such classes is that the instances of phenomena that are included in them are "equivalent," that is, they are indeed cases of the same thing in terms of the criteria that have been set up for admission into one or other of the investigator's categories. However, "suicides" in the way that Durkheim assembled them cannot really count as proper instances of equivalence, for who knows how cases of suicide are identified by coroners, who can possibly suppose—given the variability in legal definitions and considerations, in the location of coroner's offices within the administrative structures of states, in terms of the organization and staffing of those offices, and in terms of the working practices of the coroner and those who investigate instances and collect data for the coroner—that cases of death are being classified by the same criteria when the manner in which they are classified must depend on so many considerations that we know nothing about. Durkheim's study is then not a study of suicides but of suicides identified by coroners, and these two are not quite the same phenomena. Many sociologists will insist that the substitution of the latter for the former is a reasonable thing to do: after all, if you want to study suicides, then what else can you possibly do? You can assume there is some sort of rough correspondence between actual suicides and the coroner's count. However, this highlights the point about the problem being that of objectification. One just cannot objectify the phenomenon of suicide itself, except perhaps in a few rare cases where someone commits suicide before the television cameras, but you propose to study it nonetheless. However, there is also another problem. Even if you have a television record of someone committing suicide, or apparently doing so, does that provide an objectified instance?

Well, is not the trouble that "suicide" is not a sociological category, set up by the sociological researcher to identify a phenomenon that they have found? Rather, it is a "vernacular" term taken over from the society itself, and one whose definitive application is not settled by a researcher's decision, but falls to a professional, the coroner, who must make that decision subject to legal and administrative regulations that define what a correct decision is. Having apparently witnessed someone's suicide, one has to await the coroner's investigations for a definitive verdict (since the suicidal nature of the death depends not only on the witnessed event but also on other considerations that may only be brought out by investigation).

What you are *really* studying, then, are the outcomes of the coroner's work, but without any knowledge of how the coroner's work went into the determination of the death categories into which specific cases were inserted.

Many sociologists take the practical view. If we are going to study anything and if we try to take the task of objectification seriously, then we may end up being able to study hardly anything at all. But ethnomethodology rather takes this view: important sociological thinkers insist that what makes sociology worthwhile is that it involves a principled solution to problems about how we understand and find out about the society. Ethnomethodology supposes that there is not really much point in setting yourself out as being thoroughly principled if you are going to have to abandon the principles all the time because they prove too tough. This does not mean giving up on the principled approach, but accepting its consequences.

Think of it this way. You propose to study suicides "scientifically" but you cannot observe or otherwise objectify the phenomenon of an individual taking his or her life suicidal. But you want to study suicide, so you are going to have to adopt some other approach, adopting a "near substitute" perhaps and studying "suicide as found in coroner's statistics." Another way is to suppose that sociology at least does not need to study suicide. Indeed, it is something of a peculiarity of sociology that it elects topics that it cannot actually access. Stephen Cole, an important figure in American sociology today, and by no means an ethnomethodologist, has edited a book called *What is Wrong With Sociology* (Cole, 2001). One of its contributors, James P. Davis, with Cole's endorsement, holds that sociology is prone to engage in undoable studies. And this is just the point that we are currently making—that the determination to study suicide is in the principled sense that requires objectification of its phenomena, undoable. Why not simply go the other way, and identify *doable* studies? Why continue insisting on studying phenomena when you know you cannot access the data that will objectify that phenomenon? Instead, see what data you can collect, and what types of social activities and phenomena those materials objectify.

This is ethnomethodology's approach. Thus, you *can* investigate the work of a coroner's office. You *can* collect data on the way in which the coroner's staff goes about its work of investigating deaths referred to it and the ways in which the organization of the office contributes to the decision as to the classification of individual deaths.

Sociological thought is not such that the study of one topic rather than another is somehow necessary, will, by virtue of the topic's nature, ensure that its investigation will mark a major advance in the discipline. You might as well, therefore, study one topic as well as another, and you might as well, therefore, study phenomena that can be observed in real time, rather than imaginatively reconstructed from whatever records you can find of things that are done and over, and themselves forever irrecoverable.

Ethnomethodology had some affinities with SI but was also dissatisfied with the interactionist's failure to take their line of thought further, and to confront more squarely problems of evidence and inference in sociology. Ethnomethodology originates in a series of very deep reflections on the principled basis of sociological thought and of the chronic problems that everyone in sociology is aware of in respect of the construction of systematically working out solutions to those problems. In some ways, it went back rather than forward, trying to diagnose the source of these persistent problems rather than trying to advance any solution to them. Its main strategy was, in effect, to *bypass* those difficulties, and to come at the study of social life in quite a different way.

Much of the working out of ethnomethodology's root approach was a single-handed task, carried out by Garfinkel (1967), and we therefore tend to use the expressions "ethnomethodology" and "Garfinkel" interchangeably here.

The divergence with SI and with sociology more generally can be easily illustrated by a mention of the emphasis that the SI approach placed on "formal" theory and analysis. The whole gist of their approach was to aim for generalities that would draw out similarities between otherwise different types of activities. The fact that the activities their generalizations ranged over differed enormously from each other did not matter, for the point was to show that even though people were engaged in different types of activities (custodial or medicinal) and with different types of purposes (morally improving or punitive), the necessities of social relations imposed the need to behave in (formally) similar ways upon them. Thus, for the main purposes of their approach, it did not matter what those people were actually doing, whether they were running a prison, or delivering bananas by ship, or entering into a religious life withdrawn from the world. This was an oddity about social life that struck Garfinkel forcefully and not only in connection with the SI approach.

People in society go about their everyday affairs doing all sorts of things—carrying out administrative, religious, domestic, industrial, clinical, scientific, sporting, musical, magical, political, legal, and other types of affairs—and this is rather generally taken for granted by sociologists. It simply assumes that people have somehow organized the affairs of the society and the society itself and then starts to ask questions about the ways in which society is organized. It was toward this "somehow" that Garfinkel, for a variety of reasons bearing on the nature of theory and method, proposed to direct attention.

He expressed the interest clearly and in a striking way with a reference to what he sardonically called "the Howard Becker phenomenon" or in a less personalized way, the matter of the "the missing what" or, yet again, the paucity of "identifying details." Howard Becker (1993) had made studies of dance band musicians and these studies unquestionably tell you a lot about the life of jazz musicians such as the types of clothes they wear, the cigarettes they smoke, their attitudes to the audiences they are forced to play for, the troubles that the work causes for their family life, and so on. The one thing they do not tell about is the actual business that brings these people together. If one is not able to play a musical instrument, do so at a certain level of proficiency and in certain ways, and to do this in close coordination with others up on the bandstand, then one would not be a participant in such occasions at all. These musicians are making music together, but there is no sociological interest that people, who sometimes have never met each other until just before they start to play, can organize this collective activity in an intricately organized way. The "what" of their activities is entirely absent from Becker's studies.

Garfinkel also cited the question asked by a sociologist called Edward Shils at a time when sociologists were very interested in the organizational properties of "small groups." After having been presented with a discussion of the activities of juries Shils sought to show what about those activities made them a small group, To answer the question "What do juries have in common with all other small groups?" Shils asked, but is not the question "What makes them a jury?" more important than "What makes them a small group?" Shils was admonished by the other sociologists for asking the wrong question, but Garfinkel takes Shils' side on this issue. It would be one thing if sociologists could answer Shils question, could say what it is about their activities that makes a group of people into a jury doing what a jury properly should do (or not), for then they might insist that Shils' question is nonetheless not so important as the one they ask. But Garfinkel's point is that sociologists invariably cannot answer Shils' question or ones like it; they cannot say what it is about people's activities that provide the "identifying details" of it, that mark it out as the specific kind of activity that it is. Thus, somehow or another, the people playing jazz coordinate what they are doing with one another to work together to produce music of a specifically distinctice type. Furthermore, that they can do this is not a peripheral feature of their activities, but in important respects is the main business that propels what they do. Never mind that the activities of those who command ships are formally similar to the activities of those who organized monasteries or yet again to those who deliver therapy in large mental hospitals, the fact remains that caring for a shipment of bananas in transit is a very different affair from developing the religious life and, yet again, from administering the daily round of the mentally ill. If sociology aspires to understand how the social order works, then it simply cannot leave unexamined the fact that all the diverse activities of society are somehow organized and conducted in their specific, distinctive ways.

The point applies to social life generally, to the whole diversity of activities, from pushpin to poetry, but is clearly one that has central application in the study of work (given that work has such a significant place among society's affairs). The point about the "missing what" can be made about the specialty known as the "sociology of work," which has, over the past several decades, tended to be concerned with what are often called the "human relations" dimension of work, and with questions about whether workers and management get on with each other (with the so-called "human relations" school wanting to cultivate good relations between management and workers, whereas, at the other end of the spectrum, the Marxist and post-marxist approaches want to approve the idea of unavoidable employee resistance to and conflict with management: management/employee relations are essentially conflictual.) Thus, the literature is replete with portrayals of the way conflict is expressed, with the ways it is contained or repressed, and so on. However, there is nothing except the most perfunctory portrayals of how the work gets done, how the workplace operates when it is engaged in production, in administration, in maintenance, or whatever else is done. It is the work that is the "missing what" of the sociology of work.

It began to seem as if the difficulty of saying what the members of society are doing had been understated, in terms of the examination of their everyday practical activities for themselves or, as it is often put, "in their own right" and "in their own terms." Rather than just taking it for granted that people will somehow deliver an ordinary society in which they e.g. make scientific experiments, repair cars, write, edit, print and publish newspapers, prepare lunch, schedule airline flights, and all the multiplicity of ordinary things that they do, which then serves as the starting place for sociologists' theorizing and investigation it seems that it should be part of sociology's obligation to understand how that ordinary society is produced to begin with, how that "somehow." Sociology should open up that "somehow," and to pay sustained attention to the understandings on which parties to society's practical affairs organize these, and to the ways in which they organize themselves to deliver whatever outcomes are output from those affairs—the progression of pedestrians along the pavement, telephone arrangement of an outing, reading poetry, management of a school classroom within which potential student trouble is contained, the orderly arrangement of flights through air space by air traffic control, the on-schedule delivery of print materials to customers, the design and building of a model of photocopier to fill a perceived hole in the copier market, or whatever. There is no reason why work activities should be particularly interesting than any other type of activities from the point of view of these activities, or should be either more or less interesting per se; but to follow up the point about the sociological study of work as centrally featuring a "missing what," the investigation of work sites can provide a strong example.

There is no idea here of making people's work activities any more or less interesting than they seem to be to the people who do them or to professional sociologists either. The first point to be emphasized about social affairs for those who live them is that they are prone to acquire the character of overwhelming ordinariness, routine, and unremarkable. Whether the work is "interesting" relative to some sociologist's conception of what is sociologically important does not rule out the possibility of studying that work to see what the work actually is, and how it gets done. The interest is in rising to the challenge of capturing and understanding in detail how the work, whatever it is, actually gets done. "Detail" might seem a trivial matter for sociologists. If we are seeking a general understanding of people's activities, then their details do not particularly matter, and, indeed, there are so many of them that any attempt to handle them on a large scale would soon be overwhelmed in their quantity and variety. However, if we are seeking to understand how people organize their actions, then we need to note that those who are doing things cannot themselves avoid the detail, organizing their activity is a matter of dealing with specific circumstance.

For example, a diabetic, remembering to collect your medicines and foodstuffs and take them with you (and carry them with you when you drive because having them with you is a condition of your license) is surely a minor detail of the day's activities, but however irritatingly minor this task is, it needs to be done—checking that you have everything with you cannot be avoided. Similarly, a minor black mark on a printed page may be hardly noticeable to most readers, and not something that matters to them or makes any difference to the way they read the printed page, but for printers, such marks can occasion the need to do the job over again, for marks of that kind can be used as pretexts by those for whom the print work is being done to try to bargain down the price or excuse their exiting from the deal they have made with the printers. Thus, there is a recognition of *the priority of circumstances* in the organization of action, because although things-to-be-done are ordinary and are heavily routinized, they are nonetheless undertaken among circumstances that can never be entirely anticipated, over which the doers have limited control and which may turn out to be troublesome in more or less expectable ways for getting things done.

Routine activities have a significant impromptu element for those carrying them out to adapt their line of action to the specifics of the circumstances as these reveal themselves. The detail is not something that people doing these tasks can simply wish away, and therefore understanding what they do is understanding them as acting in circumstances.

Ethnomethodology was utterly unusual among sociologies in that it did not aspire to provide some general sociological theory that would abstract out the "content" of the activities that people were doing. Rather, it encouraged the development of a program of studies that would attempt to approach and capture the phenomena of social life without a stock of sociology-derived preconceptions about what people must be up to when they lead their daily lives. At the present stage of sociology, the capacity to see the ordinary affairs of the society plainly and for themselves is obstructed by all types of supposed requirements of intellectual work, and it was against these obstructions that ethnomethodology encouraged paying close and sustained attention to the most ordinary and unremarkable of doings, ones that even those who do them as the matter of their lives might not deem them worthy of attention, and most specifically to the "somehow" of those affairs: just how are those who are organizing conduct on the street, in the workplace, in the domestic setting arranging their activities. In what sequence, and in what order do people carry out the lines of action that they follow through, and how do they coordinate the things that they do with the things that others do in order to carry out joint lines of action that fit (often) closely together to produce successful acting together? In other words, the taken for granted was brought into the forefront of consideration.

Given these directives, the examination of *any* activity was pointed toward:

- The examination of activities in sequential and "real time"' terms: how is the focal activity done step by step
- The examination of the 'interactional' constitution of a course of action, considering the ways in which participants in an activity might (for example) take turns in executing steps in the course of action that carried out the activity and
- The attempt to determine the basis on which the parties were able to know *what to do next?* What kinds of understandings had to be in the possession of the parties to enable them to grasp what kind of situation they were currently in, what doings the situation called upon them for, and how to ensure (if they could) that any common understanding necessary to their joint doings was established and sustained?

The cumulative effect of these points is to bring about a reorientation (these days, ethnomethodologists often call it a "respecification") of sociological concerns, placing the central focus on the question: "What are the understandings that the parties to any social setting have, and how are their activities organized on the basis of those understandings?" It needs to be remembered that these understandings are not the participants' general ideas about how things are, their opinions about these (although these might relevantly figure), but are primarily about their practical understanding and mastery, their ability to get things done, to make ordinary things happen, and so attention is on their grasp on how an activity gets done, on what they need to do to get it done once again, and on what they and others might need, in their various ways, to do together, accepting that in many settings this practical know-how might be heavily technical in character.

The further effect is to motivate the very close inspection of whatever materials are collected from any social scene, looking at them in the most finely graded way to see what type of understandings are implicated in them, in recognition of the fact, as noted above, that the doing of practical affairs is unavoidably enmeshed in its details. Furthermore, the emphasis was on the collection of records of "naturally occurring" phenomena, and on attempting to observe and record whatever happened, regardless of ideas about whether this might be sociologically relevant or interesting. Rather than try to bring people around to talking about things that the sociological researcher might find interesting, the participants are encourage to give the lead (through their practical actions and responses) in deciding what is relevant, important, interesting, etc.

The desire for closely detailed recordings of naturally occurring activities led to the adoption (wherever possible) of audio (and, subsequently, video) recordings of activities and to the development of a powerful offshoot of ethnomethodology.

3.3 CONVERSATION ANALYSIS

Conversation analysis (mainly the creation of an associate of Garfinkel's, Harvey Sacks, though also in collaboration with Emanuel Schegloff and Gail Jefferson) has now developed into a busy, productive, and highly specialized field of research that is nowadays as much within the domain of linguistics and language studies as it is in sociology per se. Its value here is rather in its methodological import, rather than in its analytical adventures in describing the organization of ordinary conversations as such. Conversation analysis is just what its title registers: the study of ordinary conversations. The most ordinary of conversations, regardless of what these might be about, whether they involve, for example, people discussing the disappointing quality of the dinner they have at the restaurant last night problems in getting the pool cleaner to come around and clean the pool, relaying minor bits of gossip, working up to asking someone if you can borrow their car, arranging to meet the next day, and so on. The point about such materials was that they could be tape recorded and that a tape recording of them would catch a great deal of what was "going on" in the situation, especially if, as many of the recordings were, the recordings were of conversations over the telephone. The talking between the parties in such a case constituted all that they had to go on; each party can respond only to the talk of the others.

The tape recording provided a kind of "hard" data that sociologists of a scientific bent said that they aspired to. The talk was mechanically recorded and could be either listened to itself or could be very faithfully transcribed by listening very carefully to the recording. Furthermore, since the analysis of conversation would often focus on only very brief extracts, often only one or two utterances, it was easy to reproduce the relevant extracts from the transcript in a paper, thus ensuring that other analysts could have access both to the analysis offered and the data that it was based on, thereby being enabled to check whether the analysis did seem to have a close relationship to the data it claimed to dissect or not, a very close comparison of the analysis with the detail of the recording could be made. This provided the kind of "objectification" of sociological data that is purportedly an essential component of any authentically scientific venture.

The availability of the recording and transcription meant that real-time, step-by-step inspection of the course of a conversation was built up. Attention could be intensely focused on the ways in which the different contributions of the speakers to a conversation are organizationally unified into "a single conversation," the ways in which the parties to the conversation can organize their activities such that, for example, the conversation gets going, does the talking that it does, and then gets brought to an end. It is only in and through the conversation, through their own remarks, that the parties to the conversation can figure out "where we are" in this conversation, "what we are doing" at this point in the conversation, and, given that this is the point we are at, "what we need to do next" to continue and build the sequence of the conversation. One could trace in detail the ways in which the parties to the conversation coordinate their actions by figuring out what the significance of whatever the other says might be, what they should appropriately say in return. Each party has, of course, to dovetail whatever he/she says to what the others are saying, and has to fit his/her turns at talk quite specifically to the things that others have so far said, which means that the organization of each utterance in the talk, its verbal structure, can be analyzed in terms of the way it fits into the environment of surrounding utterances and contributes to the building up of the extended sequence of talk as well as, of course, the way that the talk is organized to get whatever the business of the conversation—telling a story, making a request, apologizing for an insult, etc.—might be.

Conversation analysts quickly figured that they could develop a very systematic method of examining their recordings and of analyzing the structure of the talk they were studying, which was to focus everything around the most obvious, and constraining feature of conversation, that it involves taking alternating turns at talk. To see how the utterances that speakers produce are organized around the need to generate a sequence of turns, to manage the distribution of the taking of turns, to forge connections between one turn and others, and so on, provides a most illuminating way of analyzing the detailed structure of talk, potentially of any kind, and not just that found in casual conversation.

Conversation analysis' story about how conversations are interactionally structured by those who are participating in them is not of particular interest here. Rather, the importance of conversation analysis for our consideration is much if figured out ways of being systematic, and detailed in the analysis of action sequences, enabling the most thorough fulfillment of the methodological suggestions that ethnomethodology has made.

3.4 SITUATED ACTION

The distinguishing feature of many of the fieldwork techniques associated with ethnographic fieldwork is that work is observed as it happens and as it unfolds. Thus, the participant observer who may be "shadowing" a particular individual is witness to many of the circumstances that a person orients to in the course of their undertaking their work. The analyst who uses video material can capture the fine detail of work that might elude them first time around, but which may be the very stuff of the work to those who undertake it. This distinctive cast of ethnography plays very well into another of the foundations of studies of work in HCI that originates in Lucy Suchman's *Plans and Situated Action*.

Here, Suchman (1987) argued against the predominate cognitive model of human behavior abroad in psychology and cognitive science. The principle Cartesian underpinning of this model is that human action is accounted for in terms of inner mental processes. Thus, cognitive science uses otherwise mundane terms such as "plans" as if they were causal perquisites of action. In this respect, the mind is a processing unit just like a computer, and all human action or interaction is accounted for in terms of interior mental predicates. Human action can thus be accounted for in terms of purposes, intentions, goals, motives, and—in the example that Suchman was attacking in the field of artificial intelligence—plans. To understand how people acted as they did, it has been posited by, for example, Schank and Ableson (1977), that they followed a script or a plan. Thus, their actions can be broken down into particular elements, which are then accounted for in terms of parts of the plan, and, consequently, "in order to" types of explanations are given for their actions.

Against this idea, Suchman argued that all human action takes place within a swarm of contingencies that are demonstrably oriented to by people in the conduct of their actions. It is in the light of these contingencies that people construct and engage in their social actions and in their interactions with one another. A simple but telling example is illustrated in Schegloff's analysis of the opening turns of conversation, where what one speaker might say, as part of the opening, depends on what a prior speaker has said, or what they believe a following speaker could say (Schegloff, 1986). By their very characterization as contingencies, it is just not possible to project in detail just how any one conversation may unfold in advance of the participants engaging in their social action. For example, we may well plan to tell someone about our wonderful vacation, but decide not to do that when they tell us about their recent bereavement.

In this respect, the formulation of a plan before engaging in action will not be able to take account of the contingences that develop in the course of engaging in a course of action and which decisively shape social action. In this respect, a plan cannot determine the course of someone's action. A plan can be a guide for some actions, but it will have to be applied in a context of contingencies, which it is not possible to formulate in advance of the action (Lynch, 2001). Consequently, the cognitive science treatment of plans as causal antecedents is problematic. Furthermore, there are indeed many occasions in which actions are undertaken with no plan at all. It is not that people never follow plans, but rather that plans do not automatically implement themselves and that with the types of plans that organizations provide, those who are to follow them have to do a great deal of unplanned activity in order to get the plan to work, to fit what they are doing to its demands.

On Suchman's argument, social action is then always situated within some complex of social arrangements. Studies of work within HCI have been oriented to this feature of social action in their attempts to study work as it unfolds within the actual circumstances in which it occurs. Often, this has involved reference to its organizational context (Button and Sharrock, 1997), but as work done by Christian Heath (cf. Heath and Luff, 1991) and by Chuck and Candy Goodwin (cf. Goodwin and Goodwin, 1993) have displayed, this can also involve very fine details of what a collaborator in the workplace is doing. These details can be subsequently built into the work of others who, by doing so, demonstrate their importance for the organization of work. In this regard, Goodwin and Goodwin's (1993) study of the control room in a U.S. regional airport displayed how baggage crew dispatchers would monitor the "sing-voice" of a controller in order to find, in the buzzing noise of the control room, the information they required to do their work. Hearing the voice they would attend to its content and thus ensure that baggage crews were ready and waiting when the plane docked at its landing stage.

The idea that human action is oriented to the context of its production, and that what constitutes context can be various and itself situationally dependent not only problematizes concepts such as plans or tasks, but also provides for an *analytic* mentality. That is to say, social action is not accounted for in terms of preformulated theoretical categories, but is analyzed within the situations in which it occurs in order to find, in the materials themselves, how a course of action has been put together. It is just this type of analytic mentality, which supposes that a general understanding of individual and social activity involves abstracting away all the contingent circumstances and features that affect it, that is characteristic of studies of work undertaken within HCI. Thus, not only do studies of work within HCI use materials that have been ethnographically generated, they also treated those materials as available for analysis.

* * * *

CHAPTER 4
Detailed Description

Studies of work in human–computer interaction have been used for a number of design purposes. Studies of the way in which people have to work with, and often work around, a system have been used to critique the design methodologies that stand behind a range of systems design. It is a critique of how to approach design based on an essential understanding of some of the organizing principles of how people interact and work with one another. A second interest has been to use studies of work to evaluate particular designs in use and to provide for the reiterative design of the system. However, one of the first domains of interest in HCI that was elaborated upon in studies of work is requirements analysis. This concerned the way in which studies of work can provide technology requirements of a different order of concern to those rendered through other requirements methodologies. Lastly, studies of work have been used to offer foundational reconceptualizations in HCI.

4.1 CRITIQUE

One way in which studies of work have been used within HCI is to critique existing design methodology. This has been seen with respect to the study by Bowers et el (1995) that we refered to earlier. The stark conclusion of that study was that workflow systems that are based upon abstract and formal understandings of work might encounter difficulties when they are deployed in real work settings because of the situated, ad hoc character of the organisation of work, which is revealed in its detailed study. In this respect it is not so much the design of the particular system that is questioned, but the approach behind the design. This methodology is one that construes work as a set of formal stipulatative processes (usually assuming that they are stipulated in the organization's rule books and manuals of procedure) when, as revealed through detailed study, work seldom follows closely the line laid down in the company specification, being a much more flexible affair that is adaptive to ad hoc practices and contingent operations in its accomplishment. In actual cases, there is often a complicated relationship between the formal rules and the actual course of work which is carried on in relation to contingencies, and requires all sorts of improvizations.

A strong example of this orientation to design that is partly based on conversation analysis and on her understanding of the notion of "situated action" is Suchman's (1994) critique of Winograd and Flores' *The Coordinator*. Winograd and Flores (1986) had developed this workflow system based on

particular aircraft in a controller's field of operation. The flight strip carries details about that aircraft relevant to its flight and is placed in a rack of flight strips being used by a controller. Bentley et al. make the point that the study does not so much lead to detailed systems requirements as they are conventionally thought of, but rather, how it provides what they call pointers to appropriate design decisions.

They give a number of examples. In developing a computer system to replace the manual system, an obvious advantage is that many of the manual processes can be automated. In this regard, Bentley et al. point out that it would be an easy matter to automate the placement of a new flight strip in its correct place with respect to other flight strips and that, as systems designers, this might be thought to be an obvious requirement. However, one of the observations made by the study on controllers is that the actual embodied manipulation of the flight strip into its correct position by a controller worked to focus attention on this and the other flight strips. This need to pay specific attention to the strip actually worked as a safety device because it allowed the controller to identify possible trouble early on. Automated positioning was then deemed to be inappropriate.

Again, flight strips are manually placed into different colored holders that distinguish the heading of a flight. A design decision can also easily and automatically assign the correct colored border to the electronic flight strip based on the aircraft's heading. However, manually assigning colors is a method that forces the controller to check their data and thus is again an important safety device. Bentley et al. also show that studies of work may not only ground design requirements and decisions in "the real world," which may otherwise remain obscured, but also be used to stand in for actual users in the early stages of systems design.

4.4 FOUNDATIONAL RELATIONSHIPS

In all of the above examples, there is something of a disjuncture between the study of work and the design—that is, there is a study and then some sort of design activity. This is even the case when there is a study of the implementation of a particular system, for reiteration of the design is done in the light of the study and thus follows the study. This disjuncture is reinforced when a division of labor between those who do the study and those who do the design is observed. The two elements in the division of labor, the designers and the people studying work, may be brought together in a design space, but it has tended to be the case that knowledge of the work setting gained through its study is transferred in one way or another to the designer(s).

In distinction, Dourish and Button (1998) have attempted to develop a more foundational relationship between one of the underpinnings of studies of work—ethnomethodology—and design in their proposition—"technomethodology." Fundamentally, their concern is not with the findings of particular ethnoethological studies of working settings, but with the actual analytical frame within which those studies are conducted. Similarly, they are not so much immediately concerned with the design of any particular system but with the design of systems per se.

They give an example of this foundational approach by drawing on a relationship between the notion of "abstraction" in system design and the notion of "account-able action" in ethnomethodology. Both elements are conceptually central to the respective disciplines of design and ethnomethodology. Abstraction is viewed as a fundamental tool in systems design, for example, user interface behavior, such as "file copying" or "file printing," are abstractions over the behavior of the programs they control, as are program sets of abstractions programmers manipulate to control the computer. The description account-able action in ethnomethodology shows that social actions are not just done, they are done so that they can be recognized as such. Actions are account-able; that is, it is possible to give an account of them.

Dourish and Button explore the extent to which this foundational respecification of action in ethnomethodology can be used to reflect on the idea of abstraction in design. They give an example of file copying. Many systems represent this in terms of a percentage bar that tells a user how much copying has been done in terms of a percentage of the total copying to be done. They envisage a scenario where copying is being done on a remote volume and copying fails at, for example, 40%. This can be due to several reasons, but the user has no resources to determine the actual reason in any particular case. Yet, why the failure has occured can be important for what a user does next.

Using the idea that human action is accountable (done so that it can be recognized), Dourish and Button suggest that abstractions can be developed for viewing the mechanisms at work when, in this example, copying is taking place. That is to say, copying is made to appear in such a way that it is recognizable what is being done. Thus, the system is able to provide an account of what it is doing as copying is taking place, thereby enabling users to better determine any remedial action as required.

* * * *

CHAPTER 5

Case Study

Button, Bentley, and Pycock (2001) describe InterSite, a distributed production management system for multiple production-print-factories that was designed following extensive studies of the work practices found in these organizations and reported by Bowers et al. (1995) and Button and Sharrock (1997). The study by Button and Sharrock revealed that the work of managing production involve methods for making the order of the work visible. These methods primarily involved the use of paper documents. For example, a simple forward loading board consisting of a sheet of A3 paper with print machines listed on a vertical axis and days of the week along the horizontal axis was used in many sites to assign jobs to machines within a timeframe. Another paper device was a job ticket that contained the details of the job such as its job number, its due date, the instructions for what work had to be done on it, and the like. Yet another paper form, the job flag, which included information such as the job number, its due date, and a part number, was inserted into stacks of printed material on the print shop floor.

These simple devices were used by various managers who, between them, were responsible for organizing and moving a job through its production phases, in such a way that two conditions for the organization of a print site were met: (1) keep the site at full production with all machines and operatives fully working throughout the working day; and (2) meet the deadlines agreed with a customer. They used the devices in order to make visible the orderly properties of the production process so that they could attend to any problems that might occur. For example, the print shop floor at any one time contains pallets of printed material positioned in the isles or behind machines. These pallets have inserted into them the job flags and also have the job tickets resting on top of them. Through these devices, the production manager is able to view a stacked pallet not as a newcomer to the print room floor would do, which is merely to see a pallet of printed material, but as, for example, a late job. By viewing the details of the job on the job ticket, it is possible to make determinations, such as "this job is going to be late if it is not moved onto the next production process now," and thus allows the personnel involved to take remedial action and hurry up its progress.

Another example is that when a job request is received, the administrative staff can quickly ascertain if they have spare capacity at the time the job is requested for by consulting the forward loading board. In addition, the forward loading board can be quickly consulted should it become necessary to reorder production if, for instance, a machine goes down. By looking to the forward loading board, the scheduler and the production manger can make decisions to reorganize the printing schedule for various machines in order to work around the ailing one. It is also a simple matter to know where in the production process a job is because the job ticket, which accompanies the job throughout its production phases, makes that explicit. Such a question may be asked if a customer has inquired about the progress being made on a job or should new requirements be issued for the job, or should the manger of one of the processes need to check how long it will be before the job hits him.

The devices that are used by factory production print shop managers have the following characteristics. (1) They are on public display: they are either on desks, which are in public places within the print shop, or they are placed on jobs, which are in full public view. (2) They can be consulted by anyone to check on the status of jobs for whatever reason. (3) They can be consulted at a glance: the information they contain is simple and is visually rendered, as in the case of either forward loading boards, pictorially, or the job ticket, through standard and patterned formatting. (4) They can be used collaboratively between people: they are both sites at which people involved in the management of production can collaborate and devices through which they collaborate. Thus, for example, at one site, the forward loading board lies on the scheduler's desk in the administrative area, and discussions on print production order between the scheduler and the production manager are often conducted around the forward loading board, which is also used in discussions of the production order. Thus, it becomes the means through which something about the production order can be pointed to, for instance. (5) The devices are all paper based. In sum, it is the public, shared, portable and visible character of the devices that constitutes their effectiveness in the work of managing the order of production in print factories.

This investigation into the work of managing production printing was done within the context of changes that were taking place in communication infrastructures, which meant that it had become an easy matter to move digital print files between geographically remote sites. The fact that most, although not all, print jobs come into the print factory as electronic files provided the systematic possibility of print factories working together in new ways, either coming together as coalitions to share work across the coalition or for load balancing within a company with many sites. Thus, for example, a new form of distributed print production could take place where sites could easily collaborate with one another. For instance, if one site receives an order that it cannot complete by the due date, it can now collaborate with another site to have the job printed by merely sending them the file over a network. There was then the opportunity to use the new technology for printing and new communication infrastructures to maximize printing capacities of confederations or collections of print factories.

However, a problem that the study revealed was that although it was possible to easily move files from one site to another, it was not possible to systematically select which site to collaborate with, simply because, unlike the situation with regard to single sites where the production capacity was visible, the production capacity of remote sites was not visible. Thus, the study examined a company that had more than 14 production sites, which it intended to network. However, for any job that one site needed to have printed elsewhere, the question was "Which of the other 13 sites had spare capacity?" To find out, factory managers would have to ring the other sites to ascertain if they had production holes, with all the hit-and-miss consequences attention on that process. Furthermore, sites would be unable to ascertain the production status of jobs they were responsible for and which were being produced by remote sites, again simply because the whereabouts of the job in the production cycle was not immediately visible to them.

Button et al. (2001) describe InterSite, which was designed as a solution to this problem. Importantly, InterSite was based on both the methods that managers use to order the production process and the types of tools they currently use. Thus, artifacts, such as the forward loading board and the job ticket, form the basis to their approach to supporting production management in the case of distributed production by making the order of production visible within the new distributed environment. Their solution uses the World Wide Web to support the sharing of these artifacts with other print shops and thereby allows visualization of the status of production across the network. The idea behind the InterSite approach is to use the network to share representations of production status in a form that is familiar to print shop personnel, to allow them to use the same methods for managing production in a distributed environment as they use in their own print shops. InterSite thus facilitates the sharing and presentation of documents to support existing human decision making, negotiation, and coordination practices. This contrasts with commercial production management systems, which see the problem as one of sharing data between distributed database systems to allow the automatic coordination of production management across different sites.

Button et al. considered an approach to supporting distributed document production that interleaves with existing production processes to broaden the options for print shop managers. Besides the options of taking a job for in-house production and rejecting the job, they added the option to accept a print job for production at a different location. Determination of whether to distribute a job is, of course, dependent on being able to locate a shop capable of producing the job to the required quality and within the required timescale. To support this, InterSite provides facilities for registered print shops—collectively, the "coalition"—to electronically "publish" their production schedules (the forward loading boards) to an InterSite server at a central location. From the InterSite home page, it is then possible for managers of print shops who participate in the coalition to access the published schedules as required to try to identify potential partners with spare capacity on appropriate machines.

The ability to share forward loading boards in a distributed environment and thus allow personnel to analyze production states of other factories in much the same way as they can analyze their own production state using their forward loading boards thus overcomes one of the major barriers to distributed production by allowing the production state of the whole network to be visible to participants within the coalition. Another major barrier is that of not knowing, once work on a job has begun, where in the production process the job is. Information on job status is required to identify potential problems, such as overloading of particular production processes, or to answer customer queries. By maintaining an awareness of the production status, shop managers can project the likely situation on the factory floor in the future and take action to mitigate problems before they occur.

Button and Sharrock (1997) had described that to maintain awareness of the production status, print shop managers spend a large part of their time walking round the factory floor checking where each job is in the production cycle. This practice relies on the properties of the job ticket which never leaves the job as it moves around the factory floor. The job ticket's location indicates the stage of the process the job is in, what has been done, what is still to be done, who is working on it, and so on. To address this, InterSite provides facilities for remote job tracking. As with the forward loading board display, the job-tracking facilities build on existing working practices in the print shop and, in particular, the properties of the artifacts—in this case, the job ticket—which are used to manage production.

InterSite provides an interface for receiving tracking information from external clients, allowing automatic routing of the electronic job folders. These clients must provide two identifiers—one corresponding to a job folder and the other to an in-tray where the job is now located. The identifiers for these objects are automatically and uniquely generated when they are first created. This then allows the connection of a variety of sensors that detect the presence of a work bag and can read the associated job identifier and send this information, along with a location identifier, to InterSite, where the electronic job will automatically be routed to the appropriate location and the status display updated. As an example, Button et al. describe how they connected a barcode reader using this interface, which can be programmed to send a specific location identifier and can read the job identifier encoded as a barcode on the job ticket. Again, a feature of single site production is pressed into service electronically in such a way that production managers are able to use the obvious visibility of the production status that their work relies on for its accomplishment.

InterSite is a complex system with numerous features that it is not possible to describe here. However, it has a lightweight impact on the methods used to do the work of managing print production and demonstrates the possibility for radical design that requires attention to the methods through which work is organized. The radicalness of this design is particularly brought out when compared to other production management systems. These emphasize the transfer of data and the outcomes of processing that data; by comparison, InterSite leaves the decision making in the hands of managers but allows them to operate within new infrastructural environments.

* * * *

CHAPTER 6

How to Conduct Ethnomethodological Studies of Work

Although a workplace, or somebody's work, is utterly familiar to the people who work there, it can—like any new location—be difficult for a stranger to understand. They are bombarded with new sights and unfamiliar comings and goings. Even if a generic setting is familiar to someone, the details of the local way of doing things can be confusing at first. This situation is particularly true when conducting studies of work and the workplace. Often, the person who is conducting the study may not have any prior understanding of how people do the things they do; they may often use terminology that is opaque to a newcomer and make references that are meaningless. How do people who are studying work go about making sense of what they observe?

Ethnomethodology's interest, in the way in which people order and organize their actions and interactions in the actual course of doing them, directs ethnomethodological investigators toward certain interests. Over the course of many investigations and with these interests in mind, it is possible to develop a number of maxims that can be deployed in an attempt to make sense of what can be observed in the workplace and how to go about making those observations. In this section, we will examine a number of maxims that have supported us in our studies of work. We will give an exposition of each maxim and provide some examples. These examples are drawn from our fieldwork notes and sometimes of those with whom we have worked in the past. They are not published materials but notes from "the field" that some published articles have been informed by.

6.1 MAXIM 1: KEEP CLOSE TO THE WORK

6.1.1 Keep Company with the Participants

Exposition. The first rule of doing ethnographic work in workplaces is to actually be with the people who are doing the work. We have seen that one of the insights that the ethnography of early anthropologists came up with was that the traditional way of understanding a culture was not necessarily the best. Traditionally, the anthropologist would find a native informant and have them tell them about the culture of the tribe. These native informants were very often people who were high in the hierarchy, a tribal chief or a village elder (people who were supposed to know what went on in their tribe). However, their descriptions of their culture were not necessarily descriptions that would be given by others from the same culture. Their descriptions were done from where they were in the hierarchy. When anthropologists started to develop the ethnographic tradition and started to live in the culture, they came to see that culture from different positions within it, and they began to find that the culture could look very different from the position in the hierarchy it was viewed from. Tribal chiefs just did not always know what was going on in the tribe; often, it would suit members lower down in the tribe to plainly keep their chief ignorant of what they may have been doing. But it was more difficult to keep the anthropologist who was living cheek by jowl with them in this state of ignorance, so the anthropologist started to learn more about the ways of a tribe than they previously did.

This may seem rather obvious when we start to think about it, but it is easy to became dependent on a native informant. You are given a quick and easy insight into a culture and have it nicely packaged and organized for you. There is an obvious parallel here with trying to understand a new organization or workplace. Often, going into the workplace can mean going in at a certain level. It will have taken negotiations and consultations to gain entry and this can often involve a sponsor. It would be easy to use such a person as a native informant. However, like tribal chiefs, it is not always the case that from where they sit in the organization, they can see just what happens in the details or the way their organization works, or they see it through the lenses of their position within the organization.

Someone who is going to study work needs to sit in different parts of the organization and illustrate the way in which other people see the organization and thus act on the basis of how they see things. This is not to say that organizations do not try to take into account their employee's views or that managers are necessarily ignorant of their organizations. It is, however, the case that one barrier that is endemic within organizational structures is information flow or the flow of the right sort of information. Very often, organizations will put in place mechanisms that stand outside of work—such as creating the position of "information officer"—in order to ensure that information does flow. However, operating that mechanism is a job of work itself, and how somebody does that job is crucial for the nature and quality of the information. How that person does his/her work is an important feature of the construction of the knowledge an organization or key people in the organization have of the organization. Yet they probably take the output of that work for granted and as a representation of the state of their organization.

Example. Ethnographic work done on "report writing" in collaboration with the European headquarters of an international company revealed some interesting practices concerning the way in which financial reports are put together. These reports are very important for the organization concerned. They give an account of the performance and the outlook for the European sector and will affect the stock price for the company.

However, writing a report is not simply a matter of gathering information and then allowing the facts to speak for themselves. Very importantly, writing a report involves a number of "writing practices" to make the gathered facts tell a certain story. Then report writers have to work out what that story is and how to organize the facts to tell the story. This involves organizational politics. Thus, at the time the field workers were involved, the European figures as they came in were not good. This would be important for how the "mother company" perceived the European organization and how the European manager and chief financial officer (CFO) would be viewed.

The report writers were thus in daily meetings with the CFO as a they crafted the story they would tell about the figures and how the figures would actually be calculated. In this regard, they used a number of practices. For example, there is no such thing as "the figures"; rather the figures can be calculated in different ways. Thus, for instance, the figures could be calculated in different ways by laying off incomes and payments, in different ways across different reporting periods. Thus, when revenue is to be counted is decided by the people doing the calculation; they base that choice on the story they wish to tell about the operation and do their calculations in such a way that it supports that story.

In this case, they were working to avoid having to say outright just how bad things were. They brought monies forward that were not actual and transferred money into certain funds to improve the picture; they laid off some costs of the organization to restructuring and looked throughout to give the very best forecasts that they could for the rest of the year.

In these terms, the picture of the organization that would be received by the U.S. mother company was one that was produced from within the work of the report writers. The report tells a story to those further up in the organization, but that story is the output of work of some groups of people, and the picture looks different to them from their position in the hierarchy.

It simply would not have been possible to understand the dimensions of the work involved, work that is consequential for understanding the organization and the flow of information in the organization without actually observing or witnessing the work going on.

The study in question was being undertaken in order to develop better support writing tools. At the beginning of the study, it was assumed that the principle problem faced by report writers was that of gathering the facts and having the facts on hand when needed. New information-seeking tools were being considered. However, after the study, it became apparent that the actual gathering of facts was not the main problem faced by report writers. It was how to interpret the facts in the best possible light with respect to certain organizational relevancies.

6.1.2 Do Not Mediate the Work Through Documents

Exposition. An observer does not necessarily understand what some work is just by watching someone doing his or her work. The people doing the work must describe it to the observer. This is best done as they are doing it so that the observer can actually find, in what they do, the descriptions they are giving them. It is then possible to observe them doing it as a routine matter. In the course of witnessing them working and also of telling you about their work, you will have all types of question to ask them that will be thrown up by them working and by their explanation.

For example, you will not understand everything they have said to you; you will see them doing things that you cannot find in their explanation and that they specifically said they did not do. To develop an understanding of what people are doing when they are doing their work, it is much more preferable for you to develop your questions about their work from within the course of the work itself. Therefore, do not use questionnaires, templates, fixed schemas, or checklists when you are undertaking studies of work, and do not have documents mediating between you and the work people are doing.

6.1.3 Work in Real Time

Exposition. Doing studies of work enables you to observe things as they happen by witnessing the work being done: work as a real-time phenomena, work as it really happens. That is, you are observing what people are doing at its pace and therefore better understand the constraints that surround peoples' work. This will allow you to understand how people are actually doing their work, an understanding than goes beyond just having it explained to you or represented to you through a checklist.

Example. During the course of fieldwork at a solicitor's office, the solicitor explained to the field worker that there was little point in spending Wednesday afternoon working with him because all he was going to do was to "run through his files." However, the field worker stayed on and witnessed the solicitor going through all of his current files, which were kept in six filing cabinets along one wall of his office. The solicitor spent each Wednesday afternoon flipping through all of the files he was working on, an average total of around 300 cases, or "matters" as they were called. The field worker noted that in the course of flipping through the files, the solicitor made notes for himself about activities he subsequently had to undertake.

This study showed that solicitors very much organize their work in the management of their matter files. Many of their matters are ones that follow a sequential process, with one step following on from another and giving rise to a particular next step. Importantly, however, a step may not be taken until the proceeding step has been made. Thus, on many of their files, there is a checklist of activities that have to be completed. However, that checklist does not announce to the solicitor that a step has been completed and, importantly, that a step has not been completed.

So, for instance, in a house purchase, a solicitor may have requested a title search from a local authority. Until that comes in, he/she cannot proceed in the drafting of the contract. At any time, two situations face the solicitor. First, the search has come back, or it has not. If the search has come back, then the solicitor will have received it in the mail. Upon opening the mail, he/she will then pull the relevant file and then sometime during that day check off that this has been received. He/she may then progress that matter; however, the only prompt that the matter needs progression is the file in a pile of files on the floor or on the desk that are in need of processing. Second, the solicitor does not receive anything, but he/she may not know that the requested document has not yet turned up because knowing that they have not received something is dependent upon their being aware that they are expecting something.

The problem that the solicitor faces is "keeping in mind" what stage all of these cases are in. In the first option he/she can do that after a fashion by having the file in view in a pile of files with its presence showing that further action is needed. However, in the second option, the solicitor has no way of knowing, other than relying upon memory that something is pending on a matter, for example, that a reply to a search has *not* been received and that it needs to be hunted up.

Thus, what the solicitor was doing on the particular Wednesday in question was what he did every Wednesday: he was flipping through the files familiarizing himself with their contents and to see if there was anything that he needed to do to progress them. For example, was there something holding the matter up that he needed to attend to, a reference, for instance, for a client from their bank had not been received and which he needed to move forward. His Wednesday afternoons were thus very elaborate and time-consuming memory joggers.

This simple observation emerged through the course of working with the solicitor and seeing what was going on as it went on, witnessing the work in the real time of its doing. It subsequently turned out that this observation was important for the development of a case management system. It was the sheer witnessing, in the real time of the Wednesday afternoon, that made the laborious and time-consuming nature of the work apparent—much more apparent than the solicitor merely saying in an interview that they checked the state of their files every Wednesday afternoon, or ticking off a box in a checklist.

In this particular instance, a new generic solution was developed around the management of case files—a case management system—that sets up a system of reminders and alerts: a digital filing cabinet.

6.1.4 Follow the Work

Exposition. Workflow is an important concept in understanding the work of an organization. Very often, workflow is represented in workflow diagrams showing the interconnections of separate processes in the workflow, their dependencies, release criteria, and the like. However, if workflow is observed as it happens, actual workflow conventions may be broken in the very name of making the flow of work happen. Also, the abstract representation of work in a workflow diagram or the specification of formal properties may miss the hidden work of making the workflow actually work in practice.

In both respects then, it is important to follow the work in order to understand where it goes and what goes on at critical junctures such as at handover points. Workflow is not just an abstraction on a page; it crucially involves people in the organization, and so it is just as important to follow people and see where they go and with whom they interact. Therefore, an important part of keeping close to the work is to follow the work and the people to see where they go and, in effect, how they go.

Example. This orientation is testified to in an ethnographic study of a global financial organization done by a former colleague of ours, Richard Harper. As a bureaucratic organization, it runs through the production of documents. Most of its economists work to produce documents about the countries they have responsibility for, so that the decision-making board of the organization can make its deliberations and produce its decisions. These documents involve multiple authoring: one economist makes his/her input and then moves it on to the next economist for his/her input, which may be affected by what the previous person has done. Thus, the production, flow, dissemination, and gathering of documents are crucial for the very workings of the organization. During the study, a new networked document management system was introduced, and the ethnographers were privileged to be able to do a short "before-and-after" study.

One of the striking observations about the "after" part of the study was that although it was now possible for the documents to be placed in a central depository and be accessed by those who needed to work on it, most economists in the organization preferred to stick to their original practice, which was to physically take the document in a printed form to their colleague who would be working on the document next. They preferred this time-consuming manual practice to the use of the new document management system. The field worker perfectly understood why this was the case.

Before the introduction of the new system, the field worker had followed both the document and the authors to observe where the work and the people involved went. Thus, the handover of documents was observed. The handover actually involved one economist taking the document when they had finished working on it and actually giving it to the next economist, i.e., literally handing it over. Thus, they did not use the internal post or place the document in a pigeonhole; they sought out the person who would be working on the document next and passed it directly to them.

During the course of this handover, the providing economist would take the opportunity to actively explain the contents to the recipient. This was not an isolated nor an unnecessary activity; it occurred routinely and served a very crucial function. A lot of the documents contained statistical information, as well as conclusions and recommendations. However, the statistics could be read in a number of ways, and the conclusions and recommendations were also not the only ones that could be made. Thus, the providing economist would use the occasion of the handover to explain why he/she believed the facts led to the particular conclusions he/she had drawn, enabling the receiving economist to understand the logic that had propelled the providing economist to write the document in that manner.

The actual hand-to-hand transfer of the document thus played an important role in the actual construction of the document by creating an interactional space that could be used by those involved to integrate their work together. Actually following the work and the people involved in order to observe where the work goes and with whom people interact in order to do their work can make transparent the realities of the workflow. In the case of the organization, the new document management system that was introduced did not create this interactional space that played such an important function in the construction of the document. Had the designers of the solution been aware of this function, it would have been an easy matter to provide a mechanism to accommodate this need.

6.1.5 Work and Organizational Structure

Exposition. Let us always remember when doing fieldwork and observing someone's work that work is always embedded in an organizational structure and within collaborative relationships. There are two consequences to this: first, work is done so as to be organizationally accountable; second, people are always doing something with or for someone else. Both of these factors have an important bearing on how people do their work.

To say that work is done to be organizationally accountable is to say that it could be done differently if it were not done within an organizational context. It is done the way it is precisely because it is being done in that context. For example, hackers do not document code, but engineers on engineering projects are supposed to do this, not because it makes the code better but because of organizational reasons not faced by a hacker, such as the code having to be understood by others in the organization. Understanding this can help an observer understand why someone is working in the way they are.

Sometimes, it is transparent that people are working with and doing things for someone else. However, at other times, it is less so because there may not be a formal working relationship between people; nevertheless, they have developed a collaboration between themselves that is mutually beneficial.

Example 1. With respect to organizational accountability, fieldwork in a printing works revealed a curious practice. The works had been granted International Organization for Standardization (ISO) quality certification, and above each of the printing machines, a copy of the processes for releasing a job had been fixed. One of these was that before an operator began the print run, a proof had to be run off and signed, as approved by the manager. The field worker observed, however, that although the operator did run off a proof, they checked it themselves and then began their print run without having the manager proof the copy.

When asked by the field worker about the discrepancy between the procedure and their practice, the operator explained that the manager's job regularly took him away from the print area and so he could not sign-off the proofs.

Printing, however, is done under a number of organizational parameters, for example, do the job as quickly as possible so as to maximize the income for the print works and to keep the delivery times agreed to with customers. Thus, the actual laying down of ink or toner on paper is done so as to be organizationally accountable in terms of time and income.

The quality processes in this case were interfering with the organizational accountability of printing. Simply, if the manager was away, he could not sign the proofs off and thus work on the job could not go ahead. The job could consequently be late, and this could impact the profitability of a job, as the customer would demand compensation for a late job.

The operators and the manager had, however, worked a way around this: the manager transferred his signatory authority to the operators when he was away from the print room and they could thus sign off their own proofs and get on with producing the print run.

In this respect, they could keep within the spirit of the ISO quality processes, but doing so displayed their orientation to the understanding of organizational imperatives, which themselves were not stated in the ISO processes and which could actually be compromised by them in these circumstances.

In so doing, the operators and the manager were doing their work so as to be accountable to the organizational context of their work. The lesson is clear: look to the organizational circumstances because people will work in such a way as to embody these even though, on first inspection, it might not be clear that they are doing this and how they are doing it. The consequences is that if a designer should necessitate a change in what might seem to be the way that just one or a few people are working to improve their efficiency, they might actually be interfering with organizational imperatives without knowing it.

Example 2. The second issue pointed to above is that people will work out ways to help each other do their jobs. For example, in another before-and-after study of the control room of one of the London Underground lines conducted by our former colleagues Christian Heath and Paul Luff, field workers observed the disruptive affect on the work of two key personnel of a new control system. Before the introduction of the new control suite, they had observed how the person who worked out the time schedules for the trains and the public address announcer used the fact that they sat close to each other to coordinate their work.

The scheduler was responsible for the ongoing development of the timetable for the line. This is contingent work in-as-much as the published schedule is constantly being disrupted by events. Examples include a train running late causing a knock on effect for other trains, a driver not turning up for work means a train cannot run, a passenger incident means trains have to be halted, a bomb scare means that trains do not stop at a station, and the inevitable small delays that put trains slightly off their running times.

The job of the announcer is also important. London Underground has developed a policy of keeping their customers updated on information that can affect their journeys and the announcer has the responsibility of keeping himself aware of, among other things, the running times of the trains and will make announcements of delays or disruptions as they develop.

The scheduler and the announcer have developed a way of working that enables the announcer to make announcements in a timely manner. As the scheduler works on the timetabling, he simply says out loud what he is doing. Thus, he produces a commentary on what he is doing. The announcer closely monitors the scheduler's commentary for information that is relevant for his work, and the field workers who captured the work of these two people on videotape were able to show just how finely they coordinated their work together; the announcer being observed to move toward the public address system before the scheduler had quite finished saying out loud a change he was making. The announcer thus displayed how closely he was attending to the activities of the scheduler. The two of them worked closely together day in day out, yet only occasionally did the scheduler say directly to the announcer something about the timetable and its consequences for the announcer, and also only occasionally did the announcer asks a direct question.

The introduction of the new control suite, however, disrupted this collaboration. Although the control suite worked well from many points of view, it required the announcer to be located in a different part of the control room from the scheduler. This meant that they could no longer work together in the manner that they had developed between themselves, and this meant that the announcer could no longer easily and simultaneously obtain the information he needed to do his job. At the time that the field workers completed their work, the scheduler and the announcer were sitting at opposite ends of the control room and were attempting to repair the problem caused by the introduction of the system by shouting across the room to each other. This was clearly not as an efficient as way of working as had existed previously and the timeliness of the announcements to the public was certainly affected.

People at work simply develop ways of working with each other in order to get their jobs done and to support the aims of the organization, which are not formally documented in a workflow and quite often are not known about by their managers. These ways of working can be very important to sustain the organization, so it is necessary to be on the lookout for them.

6.2 MAXIM 2: EXAMINE THE CORRESPONDENCE BETWEEN THE WORK AND THE SCHEME OF WORK

6.2.1 Reading Off the Procedures Is Not Sufficient for Design Purposes

Exposition. Many organizations have instructions or schemes for how work is to be carried out. Often, these are written down and are part of the processes for regulating both the organization of the work and the effort to be expended by those who do the work. These processes are often aimed at regulating the workflow in the organization. However, what the formal processes do not cover is the work that is done to make the processes work.

A process is very much like a rule. A rule will tell you how to do something, but the rule does not itself tell you how to apply it. The philosopher Wittgenstein explained this with a simple example of a signpost. The signpost tells you that the village of Upper Slaughterham in that part of the United Kingdom, known as the Coxwolds, is 1 mile from the village of Lower Slaughterham. However, the signpost does not tell you which way to follow it—should you go in the direction of the pointed end or should you go in the direction of the blunt end. You know which way to follow it, but you do not know that through reading the signpost, which itself does not tell you 'in so many words' which way to go. You know which way to go because you are a competent member of society and have a stock of cultural knowledge about the ways in which things in your society work including signposts. Thus, in order to use the signpost, you have to do some work, you have to apply your stock of cultural knowledge to the case in hand and follow the signpost in the direction of its pointed end. You know this so well that you probably never have to explicitly think about it when you follow a signpost.

Think of all those countless things you know about without even thinking about them that become so visible to you when you have to instruct children in the ways of the world.

Like a signpost, a rule does not tell you how to apply it. It does not even tell you that a particular activity that you wish to undertake falls under the auspices of that rule.

Like signposts and rules, organizational processes do not tell you how to apply them with regard to any particular activity someone may be undertaking. Like following a signpost or applying a rule, there is a lot of taken-for-granted unseen work that those who apply the process have to engage in to make the process work in the circumstances of any particular activity that is deemed to fall under the auspices of the process.

All organizations have this "hidden work" that everybody who does the work knows about and engages in to make the formal schemes of work, and the processes actually work but which are not referred to in the processes themselves. You will not know about this work from merely knowing about the organization's processes, nor will you necessarily know about it from just interviews with people. You can come to know about it by observing people working under the processes and looking for disjunctures between your understanding of the process and the way they are actually carrying out some work activity, and then asking them about it.

It is important for you to know about the organizations' hidden work in at least the following respect. Should you believe that the organizations' processes could be changed in order to make the organization more efficient and more productive, then you should understand the risks that are associated with your recommendations. Changing organizational processes may make it difficult for the hidden work to be conducted, and thus your recommendations may actually negatively impact the efficiency of the organization. Also, any system you introduce into the organization should not just support the organizations processes, or change processes; it should also support the hidden work. It is of little support to the organization if your system, upon implementation, inhibits the ability of the organizations members to get on with what they have to do in order to make the processes work.

Example. A couple of examples are useful here. During a study noted above (Bowers et al., 1995) a print works with numerous sites introduced, a new production management system. The Covalent system was an off-the-shelf system that was introduced into the print sites in order to support and regulate various stages in the production process. However, rather than supporting the organization, it universally led to systematic disruption of the production cycle. For example, at one of the outsource facilities the company ran, the work of the print room was at its worst point being delayed by 1 week. Generally, within the company, after two months, either the system was not being used and the sites had reverted back to their original systems or the system was only being partially used. The problem was obvious to the ethnographers who happened to witness this and to the people who were actually involved in doing and organizing the work of the print sites; however, it had not been obvious to the consultant the company had hired, nor was it obvious to the senior managers who had followed the advice and purchased this solution.

The problem resulted for the following reasons. The workflow of a print shop may appear to be very simple. A job goes through a number of sequential phases:

1. Receiving the job and putting it into the organizations processes (e.g., scheduling its production cycle)
2. Origination
3. If it is a wet ink job, plate making
4. Actual printing
5. Finishing
6. Dispatch

The company had a number of processes that governed when and how a job could be passed from one stage to the next. For example, an operator could not start work on a job until the job had been assigned a job number. The existence of this rule, like all the other dimensions of the processes for organizing the printing of jobs, was intended to ensure that there was a smooth flow of jobs across the production floor, that stages in the printing process were not overwhelmed while other stages were under worked, and that a proper accounting of the work done on a job was in place to assist pricing and costing.

However, on occasions, due to contingencies that inevitably arise in the printing operations, the outcome that the processes are intended to produce might be in danger of breaking down.

For example, it did happen that sometimes when electronically printing, an operator may have finished a current job but that their next job had not yet been fully processed by the administration staff and that a job number had not been assigned to it. On these occasions, either the production manager or the actual operators themselves would break the rule that a job could not be started until a job number had been assigned and actually start work on the job.

For many instances, this was perfectly possible because many of the jobs were repeat jobs, so the operators already had the file and they could start off the job. Thus, should it appear to the production manager or the operators that there was a potential disruption in the offing, they would consult with the administrators, find a job they could start off, and do so before the job number was assigned. They knowingly broke the rule.

The sites' senior managers often admonished both the production manager and the operatives for doing so, and they would probably not have done it if one of those managers had been around. The point, however, is that they exercised their judgments within the actual circumstances in which they were working in order to bring about the intended outcome of the rule and the processes governing production, the smooth flow of work across the factory floor, on occasions when the rule would not actually deliver the intended outcome. This work practice of bending the rule to fit the circumstance they were actually facing was not part of the process. It was the hidden work of making the process work. Not hidden to those who did the work, but hidden to the processes.

And even though senior managers were unhappy with the idea that the processes were being broken, they were very happy with the outcome of breaking them and would try not to look too hard.

The company had a simple but adequate way of recording work that was done. Simply, the operators kept a paper log of the work; thus, it was an easy matter of once the job number had been assigned, it could be entered in against the work that was already underway.

The system that the consultants had recommended and introduced was causing such havoc because it prevented the operators from engaging in their ad hoc practice of starting work on a job before it had been entered into the system. Under the regime of the Covalent system, an operator could simply not start work on a job if there was no job number because they were required by the system to log in their own personal operator's number against the number for the job they were working on, and unless the system had both it would not admit the job into the system.

Operators were also unable to follow their practice of catching up and of entering the job number into the system once it was issued because the system also recorded the time it took to print the job for the purposes of costing the job. If they pursued this practice, then jobs would be seemingly taking much less time to print than they would otherwise have done.

6.3 MAXIM 3: LOOK FOR TROUBLES GREAT AND SMALL

6.3.1 Troubles Are Instructive

Exposition. Troubles at work are very instructive with regard to understanding how the work is organized. For example, when trouble occurs, it makes visible the structures through which things are normally done.

It might help you to think about this first point with reference to everyday life. The conventions about the direction in which to drive along a one-way street will soon become apparent to a novice driver if they drive down it the wrong way. The reactions of other drivers will make the normal order of the flow of traffic apparent. It is the same with regard to organizational matters—when something goes wrong, it usually becomes plain how it should be done.

Example. In research at a solicitor's office, it soon became apparent what the normal procedures for moving files around the practice were when a file went missing.

Everything in the work of the solicitors under study was organized around paper files. It had been observed by the field worker that these files were piled on the solicitors' floors, on their desks, in the hallway, and on secretaries' desks. It was also observed by the field worker that files moved from one location to another. However, it was not until a file went missing that the fact that there was an order to the piles and the movement of files became apparent. The piles of files were not random; rather, stacks represented a sequential priority system, with ones on the top being more urgent than ones at the bottom.

Also, there was a pile that indicated "matters" for processing and where they were actually placed represented the type of processing that was required. Thus, ones that required the secretary to make an input were placed at a particular position on one of the filing cabinets, and on the secretary's desk a pile in a certain position meant that they were ready for collection by a particular solicitor and those that were placed at another position were for another solicitor. When the file went missing, it required searches of other piles to see if the solicitor or the secretary had inadvertently placed them in a wrong pile.

What to the fieldwork at first appeared to be a random, even chaotic, littering of files all around the building actually turned out to be an orderly and ordered assembly, which was understood by those who were working on it and whose purposes it served. This became apparent to the field worker in the course of the hunt for the missing file.

6.3.2 Do Not Measure Troubles According to an External Standard

Exposition. External standards are often used to measure work, for example, customer satisfaction. How well a customer is satisfied is a measure of the work. Thus, in trying to understand work, the external measure might be used to provide for the work rationales. However, although there may be an external standard, it may not be the actual measure that those who do the work value. They may, for example, value "autonomy"; thus the way they do their work involves keeping aspects of it secret from their manager to allow them certain freedoms. Thus, they may do their work to satisfy customers, but they could do that by doing it in some other way; doing it in the way they are (and satisfying customers) allows them a certain flexibility by keeping aspects of what they do concealed.

In this manner, it is possible to try to better understand the way in which work is being organized to achieve those things that matter and are valued to those involved, which may or may not be consistent with an external standard.

Example. Fieldwork conducted on a software engineering project found that the management used a problem-solving wheel, which was an external measure that the company had introduced. The problem-solving wheel required that the number of development problems be logged and progress made on solving those tracked. The objective of this tracking was to ensure that the rate of problem solutions coincided with the elapsed time of the project. Thus, at a point 90% through the project, there should only be 10% of problems to fix.

However, the management did not use the problem-solving wheel in the manner specified. They also measured progress with respect to the difficulty of the problem. This is because they "knew" that the engineers would fix the less difficult problems first, leaving the harder problems to the end. Thus, although the problem-solving wheel might show that, for example, only 10% of the problems remained when they were 90% through the project, this might be the hardest 10% of the project, which would require much more than the 10% of the project time that was left. What mattered was not the percentage of problems that was being fixed, but that the project would be completed on time. Therefore, measuring how severe the problem is and having a proper estimation of how long it would take to fix it was more important than meeting the measurement system of the problem-solving wheel.

6.3.3 How Do People Distinguish Between Normal Troubles and Major Hassles?

Exposition. All jobs have their normal troubles—troubles that are just part and parcel of doing the work—ones that do not always occur but when they do no one is surprised. How people deal with them provides an understanding of how work is organized.

Example. During a study of production color printing, it was found that there were a number of sources that gave rise to normal troubles for the printers, for example, customers, machines, and sales force.

Customers in this business were not the end client, but were design houses, and they could be relied on to cause difficulties. For one, they would promise work for a certain date but it would be late, yet they would not take into account this fact when demanding that the job be produced by the date agreed. Another problem is, because they were staffed by designers who were not trained in the ways of printing and thus knew little about the skills and effort involved in laying down ink on paper, they would produce their designs so that they looked good on the computer screen. However, these designs could cause difficulties for printing simply because of the practicalities of laying down different colored inks in various combinations with one another on paper. Thus, a job might turn out to be much more complicated to print than estimated for, simply because a designer had produced their design for the page in a certain way.

Machines routinely break down and seemingly always at the worst moments. Also, the sales force cause problems for the accounts department because they are motivated to get in sales and reach their targets regardless of who the customer is. Yet the accounts department has an understanding of "good" and "bad" customers, customers who pay on time and customers who have to be chased.

However, although these are problems that do have to be contended with, they are normal troubles: familiar and known-about problems, problems that are viewed as part and parcel of the work of color production printing, and problems that the people who have to deal with them have worked out methods for solving. Thus, for example, account executives use their past experience of customers and build in longer periods for jobs that come from certain customers, the scheduler and the production manager build in buffer jobs to contend with the possibility of machine failure, and the accounts department has developed ways of intercepting sales force order forms.

However, color production printing at the company involved, at the time of the study, was faced with two major problems that it was contending with and the way in which it was doing that required an overhaul of its business practices. These included the falling average print runs and the many new competitors that surfaced in the marketplace resulting in tighter margins. How a previously successful company, which had now gone into the red, could return to profitability was a major concern. How was the company to position itself with respect to its competitors and how was it to grow its business with technology that was designed for long production runs when production runs were systematically falling were also major questions?

Thus, people and organizations distinguish between normal and major troubles. Understanding how they overcome normal troubles allows you to better understand how they do and organize their work because part of that will involve defending against these types of troubles as part of the contingencies that surround their work. With regard to major troubles, one of the opportunities they present to the ethnographer is that he/she can determine to what extent these troubles actually present themselves in the everyday work of the organization and with how the organization's solutions are actually being oriented to and are percolating down through the organization. The ethnographer has the opportunity to provide the organization with a reality check as to how its proposed solutions are having an impact on the way the organization operates for real.

6.4 WORK IN ITS OWN TERMS

6.4.1 Not Theory-Driven

Exposition. There are a whole range of theories that have been developed in management and business sciences about the best way to organize companies, to manage change, and to support organizational processes through technology. However, at any one moment, it is usually the case that one particular theory is fashionable. For example, it was not long ago that everyone was advocating and saying they were practicing 'business process reengineering,' although now this particular focus has waned.

Studies of work of the type examined here, however, throw theories about work and organizations away. Now, this might introduce a certain apprehension in some because it is comforting to encounter unfamiliar territory with a known resource and make it familiar with a standardized theoretical framework. However, theories of work are about providing an account of the work from outside of that work, providing an explanation about influences that are said to govern work such as managerial structure, gender, economic forces, and the like. Theories of organizations and work do not tell you how work is actually done.

Example. In his very popular book, *Images Of Organizations*, Gareth Morgan (1998) outlines seven images of the organization that have been used in management theory and which have been articulated within countless businesses around the globe. One of these is the image of the organization as a "machine." This is an offshoot of Taylorism in which the organization is seen as a machine, that needs to be fine-tuned. The object of management is then to exercise this fine-tuning on what are thought of as essentially mechanical components, the people who do the work of the organization (Taylorism's idea was that all the thinking in an organization should be done by management). Thus jobs they do are to be broken down into the smallest components, so that the work involved does not require thought or decision, designed as repeatable sequences of tasks, each measured and each standardized so that people can mindlessly follow them. This makes for standardized practice through the organization that can be measured and the machine is thus tuned and oiled.

Another way of perhaps rendering this management theory is, to borrow a metaphor from Tracey Kidder, the "mushroom" theory of management—keep people in the dark and pour manure over them. However, as has been discussed in previous sections of this book, when you do ethnographic studies you will find that people are constantly working to make the organization work properly. For example, remember that processes do not apply themselves, it takes active work to apply the processes, and without that work the processes are for all practical purposes irrelevant. Going into an organization with a theory of management, or a metaphor or any theoretical heuristic, will blind someone to the way the work is really being done, and the purpose of doing studies of work practice is to make visible that work. Theories are not a good tool to help you. One of the features of using a theory is that it is often the case that the work is described in such a way that it fits the theory, which inevitably distorts the work.

6.4.2 Tell-It-Like-It-Is

Exposition. There is a tendency in professional depictions of work to render work in the jargon terms of the profession. For example, the term "workflow" is so often used that it is easily forgotten that this is a professional term used to represent certain states of affairs and often its relationship to the actual state of affairs on the ground is unknown. It is highly unlikely that people doing the work described by a workflow diagram speak about their work in that way or even understand their work within a workflow structure.

Also, consider the way in which the work of a unit is professionally analyzed in terms of value flows. For instance, one way in which organizational consultants try to understand the functions of one of the units or departments in an organization is to measure the value that flows into that department and then measure the value that flows out of it. The differences between the flows represent the department's value add and the consultants then believe they have been able to measure the work of that department in terms of its value creation function. Should there be no value add and the value out equals the value in, then you might recommend changes to the value flow or changes in the work of the department.

Thus, as a professional matter, consultants might render work in terms of "flows," "values," "roles," "dependencies," "risks," and the like. When they do this, they are really drawing off a theory of work. Thus, using the terms of a theory, the components of work are those of the theory, which may, but probably will not, be the components that people are using to build up their work. This is because the theory is a way of accounting for the work as opposed to analyzing the work. The value of doing a study of work practices is that you can analyze the work in its own terms. What are the practices that people are using to order and accomplish that work is the prime question, and often those practices are masked by professional jargon and theories.

Example. An extended example will help to understand this point. The example is intended to illustrate how the jargon term "workflow" can blind us to what is going on in the doing of the work that is glossed in that term. It is an extended example because it also illustrates what is meant when the phrase "work practice" is used in work practice analysis.

The example is taken from the field notes when undertaking the study that featured as an example used above (Bowers et al., 1995).

The print company that featured in the study had introduced, on the recommendation of their consultants, a new production management system, Covalent. The designers of the system had analyzed the workflow of print shops and described this in terms of various process stages a job went through, introduced descriptions of what were said to be release criteria, commitments, roles, and functions. However, their use of the idea of workflow blinded them to the work that went on in those processes, and they failed to understand many of the work practices that were engaged to actually make the workflow. As described above, the failure to understand these practices resulted in a system that, because it now precluded these practices, actually hampered rather than supported the flow of work in the print shops the company had introduced Covalent into.

The fieldworks found numerous practices that were not even noticed by the professional consultants.

- Juggling the in-tray

In one particular site that featured in the study, the print company worked inside a U.K. Government department that had outsourced its printing requirements to the company. A print job was triggered by the receipt of a request for a print job on forms that were circulated throughout the Department. These forms solicit details about the job such as the cost code it is to be printed under, the type of paper to be used, its color, the number of copies required, etc. Upon receiving a job request, the job is assigned a job number that usually corresponds to the number on the request form, and the request form is then used as a docket that accompanies the work and enables operators to see what they should be doing now and what should be done next. The administrative staff place the docket and the work to be printed into a transparent jacket and assign it to an operator by physically placing it in the operator's in-tray, which is located within their working area. The company operates a 10-day turnaround for jobs and the printing order is organized on a first-come, first-served basis. Thus, operators are supposed to order the printing of the jobs they have been allocated by working up from the bottom of the pile in their in-trays; in short, there is an intended sequential order to the printing.

However, operators do not necessarily process the work in their in-trays in this sequential fashion if, in their judgment, the sequential order cannot ensure a smooth flow of work across their machines. Instead, they may juggle the contents of their in-trays, reordering the jobs so that they can smoothly pace the flow of their work. In reordering the in-tray, operators will examine the job dockets in order to make assessments about how complicated the job is, or how long it will take them, and balance these assessments against the deadline for the job, their current workload, their need to print jobs that have not been completed from the previous day, the fact that there may be jobs that they know will be upcoming but which they have not yet received such as regular jobs, and the like. With this information in hand, operators reorder the jobs in their in-trays in such a way as to best optimize the scheduling of the work-in-hand to ensure a smooth flow of work.

An example may help to clarify this practice. A job requiring a long print run will tie up a machine, possibly for a number of days at a time, and during this period, the operator may have little to do other than monitor the paper supply, paper jams, print quality, and the like. Under these conditions, the use made of the machine is optimized, whereas the time of the operator is not. However, the opposite condition can occur when there are numerous short jobs to do, for the operator may be hard-pressed for time while the machine is not being optimized. These extreme situations arise because the machines they use are able to do two jobs at the same time: scan in jobs to be printed while at the same time printing other jobs. Scanning in jobs allows operators to perform various editing operations such as masking blemishes, cropping and sizing, adjusting skew, merging documents, and the like. These operations are labor-intensive especially so if the quality of the original hard copy is poor, and while scanning in images, an operator may not be printing, thus not maximizing the print capacity of the machine. Hence, due to the design of the machines, two extreme conditions can develop: one where the operator is idle and the machine is working, or the other where the machine is idle and the operator is working.

Operators are able to use another feature of the machines, however, to minimize the possibility of these extremes developing and around which they may juggle their in-trays. This feature is that the machines are also able to store scanned-in jobs in a "job queue," and thus one way in which an operator may juggle the in-tray involves setting off a long job well in advance of its completion date and not in the sequential order it was placed in by the administrative staff in the operator's in-tray, so as to give themselves time to scan in a number of short jobs that can then be printed later on. Indeed, these may be printed as if they were one long job during which time operators are able to scan in other jobs. In this way, operators can maximize the use of the printer and their own time, ensuring that they are not idle and ensuring that the printer is in constant operation. Thus, operators juggle the contents of their in-trays, constantly reordering the jobs so as to achieve an order to their printing activities that ensures a smooth flow of work across their machines.

Even so, they can have holes in their production schedules or find themselves overstretched, and they have devised other methods and engaged in other practices to cover these contingencies.

- Jumping the gun

Operators can use work that they know is upcoming, such as regular requests, as a fixed point reference around which to schedule other work; however, although a useful resource, few of the job requests received by the company have this known-in-advance character. To maximize the benefits for scheduling afforded by fixed-point references, operators engage in practices for temporally investing work with a know-in-advance character and thus achieve for some work a fixed-point status around which they can juggle their other work. One of these practices is to anticipate a job, which they then use in ordering the schedule of their other work. Anticipating a job involves "jumping the gun" and beginning work on that job prior to it being entered into the production cycle by the administration staff. Not all work lends itself to being so treated, print-on-demand work being a notable exception.

By their very nature, print-on-demand jobs do not require setting up; they merely involve producing a specified number of copies of already electronically stored work. Print-on-demand job requests are also received in a different way to other job requests, which are either hand-delivered or come through the internal and external mail systems. Although these requests consist of an order form and the original hard, or electronic, copy, print-on-demand job requests do not have to be accompanied by the originals because they have already been set up and stored on the print machine's hard disk. Also, because print-on-demand requests are often immediate responses to a request made by the department's own customers, they are typically rung through to the administrative staff who will record the request on a memo pad rather than the job request form that accompanies the originals. Furthermore, because these requests are rung in, they do not follow the normal pattern of entry into they system, which is typically done at the beginning of the day when the mail is processed. The print-on-demand memos will thus normally lie on the administrator's desk until the following morning when they are entered into the processing cycle along with the job requests that have come through the postal systems. Thus, at any time of the day, a number of white print-on-demand memos may be lying on the desks of the administrative staff awaiting entry into the system.

Operators are able to use these features of print-on-demand jobs to invest them with the temporary character of known-in-advance work and then use this character to address scheduling anomalies. Knowing that they have a print-on-demand job in the offing, they can juggle their other work around it. For example, if their pacing of the work suggests that a production lull may be developing, operators may begin work on the print-on-demand job, even though they have not yet been assigned to it by the administrative staff, and thus use it to plug a production lull. Alternatively, they may want to get it out of the way in order to allow themselves more time for a job in their in-tray that looks complicated and which, if they were suddenly faced with a print-on-demand job, would be problematic to produce. Consequently, operators will call in on the administrative staff in order to acquaint themselves with the deposition of any white print-on-demand requests and possibly "jump the gun" by loading in the electronic file and printing it off before they have received the job docket. Jumping the gun is then a practice that operators engage in as a way of creating, for themselves, a fixed point of reference by investing a job with a know-in-advance status, which they may use to, in part, achieve the smooth flow of work by plugging production lulls or getting work out of the way.

- Monitoring each other's work

The print rooms at the sites are arranged so that similar types of printing machines are clustered together. For example, at the Sheffield site, the DocuTechs (the digital copiers/printers), face each other, as do photocopiers. Even the two photocopiers that make up the counter service are placed adjacent to one another, although they are separated from the rest of the print room by a wall with a large window in it. This configuration of the machines means that operators have a clear field of vision of what other operators are doing and are thus able to monitor whether an operator is working on or is absent from their machine while it is printing. Operators can also monitor whether one of their number who is actually present is also attending or not attending to the printing process because they are in discussions with the manager, other printers, or administrative staff.

In addition to the resource provided by direct line of sight, operators have other ways of monitoring each other's work. First, the machines emit warning noises that alert operators to completed processes or machine problems such as paper jams or depleted paper trays. In addition to 'designed-in' sounds, the machines also make regular noises when engaged in particular operations and skilled operators can monitor the printing process just by the sound that a machine is making at any one time. The neophyte to the print room encounters a cacophony of undifferentiated sounds, merged together into a booming, sometimes overwhelming noise. To experienced print workers, however, the noise can be disassembled into its component sound parts, and they can draw out a sound from the noise. Thus, what is mere noise to a neophyte is a rich resource of meaningful sounds that can be used to monitor the overall work of the print room and the state of particular machines by the experienced operator.

Operators also monitor each others' work by keeping abreast of each others' workload, which they do through various means. First, they may draw off each other's experiences, consulting an operator who has done a similar job, if they require advice, or working out with another operator on how best to set up a new job and thus, in the course of seeking assistance, disseminate knowledge of their work around the print room. Second, regular jobs are usually done by the same person at the same time each week, and other operators know who does what regular job and when. Third, they familiarize themselves with each others' in-trays. Fourth, they will discuss with the administration the correctness of assigning a particular job to themselves or to another operator. Fifth, they enter the administration office and see new jobs that have come in such as print-on-demand jobs.

In summary, print operators engage in a number of activities through which they are able to monitor the workload of other operators and themselves. The upshot of this is that operators are able to make a realistic assessment of the work of the print room and the work of other operators so that at any time during the day they will "know" what each other is or should be doing. If in doubt, they can always check because each operator keeps a list of jobs completed and notes on the course of a job on their machines, which together with their in-trays is available for inspection.

Consequently, operators are not only able to monitor each others' work but do so in some considerably fine detail. They put this knowledge to work in order to, in part, achieve for and with each other the smooth flow of work through the print room by intervening in each others' work at decisive moments should they deem it necessary. Thus, for example, they are able to use their ability to "see" and "hear" the state of a machine and to see the presence and attentiveness of another operator to keep a job running for them should that person be away from or disattending the machine when the paper tray needs to be filled or should the machine "go down", or should there be a paper jam. They use their ability to see and hear a machine and to tell when the job that another operator is running is complete. They also use their ability to see the presence and attentiveness of the operator and their knowledge of the operator's workload to start another job off for them and to move their work onto the next process should that be relevant.

Although on the face of it the print room is staffed by a collection of individual operators with responsibilities for their machines and their own workloads, the operators regularly engage in activities that testify to the fact that they are, as an utter feature of their routine work of printing, monitoring and attending to each others' work. These activities are ones that are designed to ensure the smooth flow of work through the print room as a whole, and across each others' machines; thus, the ordering of the flow of work is, in part, achieved through the mutually supportive actions of the operators designed to handle the contingencies of operator presence and attentiveness and machine state.

- Passing the work to Mary

In assigning jobs to particular operators, the administrative staff makes decisions about what machines are best suited to a particular job, the skill of individual operators, what jobs operators are currently working on, the urgency of the job, etc. The outcome of these decisions is that the work of the print room as a whole is formally ordered. What we mean is that the organization of the work is accountable to management and that the assignment of work to operators is considered to be the product of rational decision making because it is done with overall knowledge of the workload on the print room.

The formally constituted order to the work of the print room does not, however, take account of the contingencies of print production as they unfold for individual operators. These contingencies can take the form of "jinxes" that plague an individual operator during the course of a working day. Jinxes can take on a variety of forms. First, an operator may be faced with a machine that "plays up" that day. That is, although the machine may not be faulty, it is not working to par and thus the operator is unable to produce as many copies in a day as they usually can—operator speed is a factor that will have been taken account of when assigning work. Second, the machine runs faster or slower according to the amount of work that has been stored on its hard disk. Thus, if the print queue is full because operators have been loading in work while other jobs have been printing, the machine will work at a "snail's pace" when compared to its empty state. Third, the specifications for the job can affect the production of the job in unanticipated ways, and jobs that may seem quite straightforward can become problematic. Fourth, operators can make mistakes; they may misread or misunderstand instructions and have to do a job again. Fifth, the tools for the job may go astray, for example, the stock control can go awry so that operators may not have the right paper at the right time, or the suppliers have mislabeled stock or be late in delivering it. Sixth, although, in the main, the machines are very reliable, they do suffer from breakdowns, which cannot be planned for.

Consequently, although the formal organization of work for the day may be put in place in the assignment of work to operators, the contingencies of the unfolding production of print during the course of a day's work can conspire to thwart the planning and decision making. The contingencies can result in the very situation that management and operators attempt to avoid: idle or overworked operators.

Operators have developed practices for handling these contingencies and achieving the outcome that the formally constituted organization of the day's work is intended to result in: the smooth flow of work through the print room. One of these involves passing over work to their colleagues who fold it into their own work and then pass it back. Thus, although the work has been done by another operator, it is done as though it was done by the operator assigned the work.

The operator of the counter service at the Sheffield site is the most flexible and the most often used in this respect. This is because the counter service is a potential buffer in the printing system. The counter service provides a day turn-around service for small jobs that are brought down to the print room by members of the Department and by the very character of this service the workload cannot be planned for. There may be slack days, but there are also slack periods at regular times each day, especially the first few hours of the day, around lunchtime, and from the middle of the afternoon onward. At the Sheffield site, the company has two machines that service counter work operated by one full-time operator, Mary, but who can be assisted by one of the dispatch workers should there be a particularly busy period. Mary is thus easily able to accommodate counter work, which has built-in slack. To take up this slack, administrators assign small, non-essential print-room jobs to Mary, which can be picked up and put down, depending on the demand on the counter.

Operators in the print room have developed a way of handling the contingencies of print production that can defeat the structure of their working day, which consists of "passing the work to Mary." Because they are within sight of Mary and the counter through the large window in the wall that separates the counter service from the print room, they can tell how busy she is, and because, by its nature, Mary can put down the work assigned to her by the administrative staff because it is work designed for her to be able to put down, Mary can take some of the work that operators cannot complete or are being slow in completing. Not all jobs can be passed to Mary because her machines may not be capable of processing them, or she may be very busy, but passing it to Mary is a practice that can be deployed as her working circumstances permit and one that is designed to reachieve the structure of the work for the day. Because operators are able to monitor each others' work, everyone knows when one of them is having difficulties and at times a mere raising of the eyes to Mary results in an agreeing nod and the work is quickly passed over. However, Mary is not the only person the work can be passed on to, for other operators will try to make space within the framework of their own day's work to fold in a colleague's work and then pass it back—'passing the work to Mary' personifies then the practice of folding in another persons work into one's own schedule of work and then passing it back.

If the work is described in its own terms, that is, just what people are doing as they understand it, then it may be possible to observe aspects of that work that are not made available through the terms of a theory. If the systems designers had a description of the work that "told it as it is," then it is highly unlikely that they would have designed a system that precluded the deployment of these practices. As noted above, the system that was introduced did this with extremely disruptive consequences for the work of the print shop.

6.5 MAXIM 4: TAKE THE LEAD FROM THOSE WHO KNOW THE WORK

Exposition. Organizational consultants are often brought into organizations as outside experts. People engaged in work practice studies, however, reverse this role and understand that the people doing the work are the local experts. You need to learn from them about how to do their work, rather than you teaching them how to do it better. This is because they work day in and day out in that locality. They build up a stock of knowledge about their company, their fellow workers, the machines they use, and the like. They use this stock of cultural knowledge in doing their work. If you do not understand how their local expertise is an integral feature of the way they work, then you may well not understand important facets of their work.

Example. Another example drawn from the study of color production printing will illustrate this point. During the fieldwork for this study, a company selling a scheduling tool for the print industry demonstrated its latest offering. The work of the scheduler in a print factory is crucial to the effective operation of a site, ensuring that machines and personnel are working to capacity, that there are no log jams in the process, and that the job arrives with the customer on time. In large print factories, this can mean juggling many jobs, machines, people, customers, and outsourcers against timelines. Automating this or partially automating the system as it has been done for many other production industries might seem an obvious move. The technology would allow schedulers to block machines against client jobs, experiment with optimal loading, group similar jobs together, and other such functions as well as interfacing directly with processes to have a real-time presentation of the number of copies printed against the number to print. However, despite its sophistication, it was not purchased.

This is because scheduling decisions were based not only on the factors handled by the system but also in the site scheduler's local knowledge about factors that could seriously affect printing, for instance, his local knowledge about customers, machines, and personnel. Thus, from experience, the scheduler knows that some particular customers are often late with their jobs, so he would take that into account when scheduling them. Also, the scheduler knows that some machines are more prone to break down than others, and that some press teams are able to be more productive than others. These matters and knowledge about other local matters constituted the scheduler as a local expert whose knowledge about scheduling for his organization would exceed anyone's from the outside. In fact, the field workers observed that he used a complex production calculus that was not only based on questions to do with print runs, types of jobs, and other objective features of a job but also on very local matters. Into the production calculus went the following (taken from Button and Sharrock, 1997):

1. The suitability, ease, difficulty, cost, and speed with which different types of work can be done on different machines or cannot be done on them at all.
2. The difference one operator or another working a given machine can make, if any.
3. Figuring out how many sheets of paper, with what proportions of wastage, running for how long on how many machines, will need to be printed.
4. What allowances for preparations, down time, delays, stock availability must be made for a particular job.
5. The cost to the organization of the print run and of the direction of economic advantage—to the firm or to the customer—that may result from any print decision.
6. The flexibility of the charging to the customer relative to the cost to the organization, as well as to the appropriateness (and of the advantage to the organization of bringing in the work, keeping a machine running, keeping a customer sweet) of charging this customer with significantly more than, no more than, or even a bit less than the cost to the organization might be.
7. A familiarization with the routines and flows of work of this organizations and others with which it might deal so as to address issues of keeping up the amount of work coming in, being able to anticipate the quantities of work in hand and to estimate (a) whether this be enough or more than enough in quantity and suitable in variety to keep all (or all the important) machines running and (b) whether, being enough to keep the machines running, it is manageable in both quantity and character with respect to getting the job done in the turn round time.
8. The operation of the in-process control of the work through the use of on-site devices, such as the forward loading board to maintain a comprehension of the current work across the site, to monitor the continuing progression of the provisionally structured succession of jobs through the sequence of print operations and through the various sections of the site, so as to estimate the progression of work on the shop floor, while anticipating exigencies that can delay the progression, and searching for opportunities to progress the work, and also anticipating the arrival of new, unexpected, or emergency work, and calculating how, if possible, to retain flexibility in the current distribution of work so as to permit the incorporation of unscheduled or additional work into the flow of work, or to allow the rearrangement of work in ways that accommodate the priority or urgency of particular tasks,

while also preserving the twin imperatives, of keeping the machines running and meeting the deadlines.

9. All of which is done through a reliance on understandings about the variable dependability (with respect to either type or specific case), with which production processes can be run, arrangements struck and adhered to, operatives to deploy the appropriate and effective skills for a job's requirements, the extent to which such requirements demand "tricky" printing operations, and the risks to which all of the above are subject

However, it is not the case that this scheduler was unique; this is in fact the work of scheduling, what someone has to do and has to know for their organization in order to schedule jobs. It would just not be possible to factor all this knowledge into a system, and the printers concerned could see that the technology they were being offered would not work for them because it was just too simplistic.

This is not to say that they could not use technological support in their scheduling activities, but that the type of support might be better designed to be closer to their work. In this particular case, it might have been better if the proposed solution had not been the automation of the scheduler's job, but rather a tool that would support his deployment of his local expertise.

6.6 MAXIM 5: WHERE IS THE WORK DONE?

6.6.1 Making Context More Telling

Exposition. The work site is the major field of operation for work practice studies. Work sites may vary: they may be an office, a desk, a customer's desk, a shop floor, even around the coffee machine. Do not let exact definitions of the workplace dictate to you; the point is that where the work is done is the place someone involved in studying work needs to be in order to understand how the work is done. You will not understand it as well as you otherwise would by locating yourself some place, interviewing people about their jobs, issuing questionnaires, or having other people describe it for you.

The reason for this is as we have seen that all work is situated. That is to say, it takes place within a swarm of contingencies. The types of considerations in the example of the scheduler above are contingencies within which scheduling is done. How people work is, in part, constituted in how they are handling these contingencies; therefore, if you want to understand their work, you need to observe it occurring within the midst of the contingencies that shape it. One of the problems of producing representations of work as in, for example, workflow diagrams, is that they tend to strip work of its context. Yet, contextual features of work are constantly making their presence felt on the very doing of the work. However, what is meant by context?

Importantly, context does not necessarily mean "within these walls." Context can involve many matters and can be derived from many sources, some of which lie outside of the particular organization. For example, they can be cultural matters. The French tend to break for their lunch and close down their organizations for the lunch period. This makes for difficulties when working across cultural barriers: if you need to contact someone in France in order to progress some mutual activity, the lesson is do not do so between 12:00 and 14:00 hours French time.

So, although you need to be on the lookout for the way in which the immediate context is affecting the work, you also need to be on the lookout for the way in which work may be done so as to be accountable to contexts outside of the immediate organization.

Example. For example, the print industry is heavily unionized and there are union rules that have been developed in negotiation with the print owners. In the two extended studies of printing that we undertook, the influence of this as a context of work was witnessed.

In one of the studies of a medium-sized high-quality private print company, the company outsourced work from the Government and this involved *secure* printing. When a secure print job was being done, it was part of the contract that the print premises be kept closed with all doors locked. It is also the case that because of the competitive position of the printers in the print industry that jobs are very keenly priced and the cost of production is closely monitored in order to ensure a maximum return on jobs. We noticed one day, however, that a secure print job was being done with the doors open and that the administrative manager was clearly in a foul mood.

The particular day in question was a very hot summer's day and it was customary for the doors of the print room to be opened to maximize the airflow. There was also a national agreement between the union and the print employers that when the temperature in a print room reached a certain temperature, the print crews could stand down for 10 minutes in every 30 minutes. Normally, when the temperatures rose, opening the doors (including the very large delivery and dispatch doors) would keep the temperature down below the cutoff point. However, because there was a secure print job being done, the doors had to be closed. The temperatures then rose to above the agreed temperature. The local union representatives then invoked the temperature agreement.

In this case, the context for what they were doing is the agreements struck between unions and employer representatives at a national level that the local workforce was invoking. It was clear to the managers involved that the job could not be produced in the timeframe agreed with the customer if work was constantly interrupted. The union representatives had offered to forego the breaks, however, if the workforce were given extra money that day to compensate them; this, however, would drive up the cost of the job and make the profit margins less, which had infuriated the administrative manager. One of the managers, however, had the idea that if they posted security guards on the doors, then they could be opened and the printing would be deemed to be secure and the temperature brought down, allowing printing to go on as normal.

It would not really be possible to understand all of the ins and outs of that working day, which surfaced obvious antagonisms between the managers and the printers. The field worker who had been on site for some time had not witnessed these antagonisms before. Without understanding the general context of the print industry, which went far beyond the walls of the print factory in question, the issue of working to rule and the import this has for the organization of the work would not have been understood. The simple lesson is that the context of work is important for understanding not only how work is done but also what context is. It is not a simple matter, and you need to be aware of all the potential contexts within which the work you are interested in is situated.

6.6.2 Setting the Context

Exposition. The above description of the production problems exemplifies that the context of nationally agreed working structures and the history of employment relationships in the print industry were a context under which the work was organized for the particular case referred to. However, it hardly surfaced as a relevant context on other occasions.

So how does the field worker settle the issue of relevant context? Simply, the field worker should not settle the question of context. It is a question that those who are doing the work will instruct the field worker in how to answer. By "instruct," it is meant here that it will be possible to see in how people are doing their work, just how they are invoking context. That is, people display how they are orienting to context in doing their work.

Example. For example, there are some theories of work that describe how workers will resist surveillance technologies. However, in a series of studies that examined the way people work in environments that involve surveillance technologies, it was found that if they view them as legitimate, rather than resisting them, they actually embrace them and indeed can find new uses for them. For instance, field workers at a nuclear power station in the United Kingdom found that the people who worked in the stores, although first weary of the introduction of Close Circuit Television cameras into the stores area, became enthusiastic about them.

While the power station was a 24-hour operation, the stores department did not have a third shift. However, the night shift needed access to the stores even if there were no stores' personnel, and so the store was left open so that night shift workers could obtain what they needed to do their work. The night shift workers were supposed to return any tools they used at the end of their shift. However, regularly, some tools were not returned, not because they were stolen but because the night shift workers just did not return them for reasons to do with forgetfulness, wishing to make a speedy exit at the end of the shift or pure laziness. This made the work of the stores' personnel difficult for they had to go and hunt up the tools from around the station, which was time-consuming and laborious.

However, with the introduction of CCTV cameras, the stores personnel could now view the night's tape if in the morning some tools had not been returned, and consequently they could view who had taken them out. Thus, although the cameras had been introduced by management into the stores (and other areas) to monitor the personnel who worked there, after a period of initial resentment by the stores' personnel, it became a legitimate technology in their eyes. They knew they had nothing to hide from management, so it became irrelevant to them as a surveillance device for how they were working as indeed it could enable management to see how well they worked. However, it actually became for them a useful tool for doing their work although in ways that they and the management had not foreseen.

We have described the idea of "definition of the situation" above, which observes that how people define their situation can be consequential for how they act within that situation. The stores' personnel defined the context of working with the technology in a certain way and this led to them work in a particular way. Had this not been the case and had they defined the context as hostile to them, intrusive, or the like, then how they worked with the technology may have been very different. If you look to the ways people are doing things, you will find out how they view the context and define it and how they build that definition into the doing of their work.

6.6.3 Where Is This Work in the Division of Labor?

Exposition. All work in an organizational context is interdependent. Therefore, a question you can ask yourself, in order to understand the work, is where in the division of labor this person or department fits. There are dependencies that exist between people and departments and they come to rely on these to get their work done. Sometimes, these dependencies can be built into a formal process for doing some work. Thus, for example, in the case of printing works, the printers come to rely on the fact that the paper stock controller will place the correct amount and type of paper for a new job at the back of their machines before setting the job up, that the correct inks will also be stacked up, and that the plate makers will have done their job and the plates will be ready and hanging in the correct positions. All of this is managed by a number of processes, so that as a job is scheduled, it triggers off requirements from various departments and people to ensure that all is on hand for the time of printing.

However, people also work in ways that support each other outside of these formal parameters and you will find that in just how someone does their job there is an orientation to doing it to make someone else's life easier. Thus, people know they are in a division of labor and that what they do has consequences for other people. Often, there are a number of ways that people can work to achieve the object of their work, and you will find that they are working in a particular way because that way they will make the work of someone else easier to do. Understanding this will allow you to understand just why someone may be doing something in a certain way even though it may appear to be irrelevant for what you understand their task in hand to be.

Example. For example, in a study of air traffic controllers it was observed how controllers hand over planes from their sector to controllers who have responsibility for the next sector. Should a controller notice that the load on his sector is building, which may have consequences for the controller in the next sector, he/she may well take actions to route planes in such a way that they do not just pass from his/her sector to the next, but do so in such a way that will make the life of the receiving controller easier. Thus, having planes cross over at certain heights with particular time intervals between them will all make the work of the next controller easier; he/she may be able to leave some planes as they are and not worry about them or divert them because they have been appropriately placed for his/her building array.

Therefore, to understand somebody's work, you need to realize that there are a number of ways in which they could probably do it, but just how they are doing it may well be oriented to others in the division of labor such that it supports the work of these others. People come to rely on people doing their work in this manner.

6.6.4 A Working Division of Labor

Exposition. It was described in the section above how all work is interdependent. This has a consequence for the very idea of what a division of labor is and how it is organized. Thus, if you inquire what someone's job is, you might well be given the description of their job as that which has been produced by human resources departments. So most organizations have job descriptions they use when they interview a candidate and furnish that for the candidate to understand their role and responsibilities. The formal schemes of work may compartmentalize and specify what the role and responsibilities not only of an individual are but also of departments. There is thus a formal view of the organization as one within which specified roles and responsibilities accruing to an individual or department interact with other formally specified roles and responsibilities in a formally organized division of labor.

However, when you start to observe people working you will often find that they are doing work that is not part of their job description and may actually be part of someone else's job description. This is because people often organize "a working division of labor" that does not necessarily correspond to the formal division of labor. You will not understand what they are doing if you hold to the formal scheme, but rather you need to understand how they have organized the division of labor for their practical circumstances.

Example. For example, in the case of one of the studies of print workers, it was seen in the practice of passing the work to Mary that operators would help one another out if one of them was hard-pressed. So when you are observing people working, always yourself ask the question "Whose work are they doing?" Is it their work or is it someone else's work? People often complain that they are doing everyone else's work but their own and that they cannot get on and do their work because of this. However, you need to remember that although their workplace and they are new to you, they are working in close proximity to people day in and day out and will have developed working relationships with them. Doing "your" work in such a way that it helps someone else do their work better or more easily, doing some of their work in doing your work because you can, and doing so will make things easier for them are all part of that informal way of working together that develops between people. People also know that others will do for them as they do for others. It is rare, although it does happen, to find in organizations that some individuals are not working in this manner. The rare exceptions bring into highlight how others work, for the rare exception is frequently remarked upon and the people in question will have a reputation for being awkward, hard to work with, difficult, and the like.

The point to emphasize though is that you need to be on the lookout for how in somebody's work they may be doing someone else's; do not rely on formal specifications of a job to describe for you what someone does.

6.6.5 Work Is an Organizational Matter

Exposition. Many people work in an organization and where they do work is an organizational matter. The way in which people work and organize themselves will be dependent on how they understand the organization. However, we should not think that an organization is merely a building or set of buildings, a name, a hierarchy, or a set of rules and procedures. Organizations are seething cultures. Hence, for example, departmental and center managers are attempting to protect their own parts of the organization, meet their targets, look good to their bosses, have rivalries with other departments or centers, and the like. Organizations are cultural complexes within which question of politics, status, prestige, and the like all vie with one another at any time, shaping what someone or some group is doing.

Example. For example, in the study of engineers that was referred to above, one of the projects was the development of a new photocopier. The software was being jointly done between a team in the United Kingdom and the United States. Each team was responsible for different parts of the software and the intention was that both parts would integrate.

It was clear to the U.K. engineers, however, that the two parts would not integrate because the two teams were being managed independently of one another and that neither team's manager wanted one team dictating what the other team would be doing. The engineers know that if they could just sit down with their American colleagues they would be able to sort the problems out. They were in e-mail contact and held conference calls but these were not proving satisfactory. The engineers know that neither of the teams' managers would pay for members of one team to fly out to meet with the other for political and financial considerations because the problems were not drastic enough. But they also knew that unless attended to, the problems would become drastic.

So the engineers stopped trying to work on the problems and intentionally allowed the problems become worse more quickly. They thus did little to try to fix them and waited for them to become drastic and drastic enough for the viability of the project to be put in jeopardy.

Neither manager wanted the project to fail because they would be held responsible for its failure, and now that the problem had become drastic, it was actually in danger of failing. The problem of integration was now one that the periodic project reviews would undoubtedly uncover, and consequently, the project managers agreed to fund representatives of one of the teams to visit and work with the other team to fix the problems. This happened and the problem was solved.

The engineers were thus working within political and economic milieu that they very well understood and which they displayed themselves as being very adept at manipulating in order to obtain what they wanted, which was a face-to-face meeting with their colleagues in order to work out the design of the software together.

Field workers studying work practices therefore need to be aware that numerous organizational matters, such an organizational politics, are taken into account in the doing of the organization's work, and they need to take this into account.

* * * *

CHAPTER 7
Making Observations

The types of features about the work and organization that Section 6 suggests are relevant for studies of work practice will not, however, just drop into your lap, you need to actively look for them. This involves the following:

1. Asking people to describe what they do.
2. Following them as they move around the organization to see where they go and whom they interact and collaborate with.
3. Following the output of their work to see where that goes and how others use what they have done.
4. Asking them questions to probe them about what they are doing as they do it.
5. Making notes of everything that occurs, not just notes about what they tell you, but who visits them, when and where, whom they visit, what they discuss with the person. Taking notes allows you to reflect later and also makes you listen. Your data are what you observe and what you are told.
6. The following are some suggestions on how to collect these data:
   (a) Open your eyes.
   (b) Nothing is trivial; do not make judgments if it is interesting or not.
   (c) It is not about you.
   (d) Be an apprentice.
   (e) Remember that people in the work site are teaching you.
   (f) There is always something going on even if it is "life is usual."
   (g) You get what you get.
   (h) There is always more, but you can get something from what you have.
   (i) A little goes a long way.
   (j) Too much time is spent collecting data rather than thinking about it.

7.1 OPEN YOUR EYES

One of the most difficult things for a fieldwork is to see what is right in front of you. This may be because it might appear trivial; for example, the person you are with may be merely "working on their computer" or merely "consulting and tidying the files in their filing cabinet." For example, as noted above, fieldwork conducted with a solicitor may reveal that every Wednesday afternoon he goes through all his current case files. He flips through them, pulls some of them out and makes some notes in his writing pad, replaces them, and continues. This may seem very trivial and not something to pay attention to. However, if you look hard at what he is doing, maybe ask him why given the amount of money he is paid he is spending unbilled time on a routine file maintenance exercise that his secretary could do, you may well find that he is in fact engaged in important activities in the management of his work as a solicitor.

You know from having worked with him that many of his activities are time-critical, that certain matters have to be done by certain times in order to meet deadlines. You also know from having worked with him that many of the cases he is working on involve input from third parties. For example, in a house conveyance, he will need authentication from the building society that the mortgage has been granted. Making these inputs can be crucial for his ability to meet the deadlines he faces. He therefore needs to know if something in the process he is following has been done. If it has, he can proceed onto the next step in the process, such as, in this example, the exchange of contracts; if it has not been done and the projected exchange date is still to be held to, he needs to contact his client and ask them to work with the building society to hurry them along.

You also know through working with him that he has 300 cases that he is working on at any one time and you may have wondered how he keeps it all in his head. Now you know how he does that. Every Wednesday, he is going through his entire case load to check that activities in the processes he is following have been done. Thus, his files have printed labels on them with various steps in the process he is following. You notice that he has some of these ticked off. So when he goes through the file he can see, for instance, that proof of ownership has been received, he has put that in the file, ticked it off, and moved the process to the next stage. He is quickly flipping through the file to see if he has actually done that, and it has not escaped his attention. However, moving onto another file he sees that proof of ownership has not been received. On checking the file, he sees that he had in fact written to the vendor's solicitors requesting that some time ago. He decides that it is time to write again reminding them and makes a note on his pad to do that.

Thus, although it is a pretty routine activity in itself, someone flipping through their files, if you open your eyes you will see that it is far from trivial but is in fact an integral part of managing his work as a solicitor, doing the things that need to be done at the correct time.

As the above suggests, if you had made a value judgment that something is not interesting, you would have missed something quite crucial in the organization of the solicitor's work. The interesting thing about fieldwork is that nothing is trivial.

If you make a judgment in advance that something is trivial, you run the risk of missing something that could turn out to be important. You may be new to the environment you are entering; if so, you will not know what is significant or not—so treat everything as potentially significant. Also, what is significant should not be judged by your standards; judge them by the standards used by those you are studying—they will become clear to you over time.

7.2 IT IS NOT ABOUT YOU

Although we are exploring work practice studies, engage in these ourselves, and are champions of ethnomethodology, it is not the case that we would therefore say that all such studies are well founded. Within the general ethnograph literature, there are many cases of 'bad' ethnography. One reason for this is that the ethnographer puts himself/herself in the narrative they are constructing; thus, they describe things not from the participants' point of view but from their own. Malinowski, who began the ethnographic tradition when his diaries were published about 50 years after his death, actually turned out to be very racially prejudiced against the people he studied.

In this respect, one of the difficult matters field workers contend with is when to make yourself scarce. Most of the time, people are very ready to cooperate with you, to have you hanging around with them, following them around, asking them questions about what they are doing, and explaining to you their activities. But working life can be stressful sometimes, and you will need to be sensitive to this and keep out of the way.

Also, some people will not be cooperative. If they are not cooperative, then get what you can and move on. For example, during one study, a manager had developed a small system for keeping track of his documents that he used in doing his work. He was convinced that if he showed a field worker how he used it, the fieldwork would pinch the idea for a system. He then had the idea that he would be prepared to show it to the field worker if he agreed to purchase it. Thus, during the time the field worker spent with him, the field worker was not really able to actually observe his work, and the best course of action was to disengage and move on.

Some people will want to be cooperative but not understand what you want. For example, in one study, the manager of a department was very happy for the field worker to work with him. However, the manager believed that what the field worker wanted was a description of what he did, the problems that he contended with, how he managed them, and the like, all of which the field worker wanted to hear about, but then the field worker also wanted to see these things happening in practice. No matter how the field worker tried to lead the manager, he could not seem to understand that the field worker just wanted to sit with him in his office as he worked on his papers, answered the phone, and the like. The field worker took what was on offer, the descriptions, and moved on.

A good metaphor for the ethnographer is that of the apprentice. Apprentices are assigned to an "old hand" in order to learn from them as they are doing their work. The "teacher" explains what they are doing as they are doing it, and instructs the apprentice in the ways of the work. Of course, the apprentice learns in other ways—for example, through formal classroom instruction—but a crucial aspect of the apprentice system is that people learn their trade within the course of witnessing its doing by experts.

Field workers need to act as the apprentice, having those you work with instruct you in their ways. In this respect, field workers need to remember that they are learning from them, not talk about how they would do things or how they have seen others doing things.

7.3 THERE IS ALWAYS SOMETHING GOING ON

Within social life in general there is no "time-out" with respect to the ways in which people are doing their affairs in orderly ways. Within the world of work there is consequently no time-out from the methodical ordering of the work activities and interactions. But it does not mean that people are actually working every minute while at work; naturally, they do other things other than work, e.g., fool around, drink coffee, steal office stationary, and conduct personal business. However, when they are working, there is no time-out from the organizing of that work as a methodical course of affairs.

This means that there is always something going on, even if it is business as usual. But business as usual takes work to make it work properly. Consequently, even if it appears to an outsider that nothing is going on, there will be something that is being done. This is not to say that what is being done will be interesting, even to those who are doing it. Any field worker will recount stories of hours spent with someone doing boring and repetitive work. However, even here, it is necessary to be careful. A field worker in an accounts department spent one whole day sitting next to one of the clerks who spent her whole day "posting" supplier invoices into the accounts system. This involved the tedious work of taking supplier invoices that were the printed outcomes of electronic account systems and rekeying all the data they contained into the companies electronic account system. This involved not just the data entry but information about the supplier. The result of this period of observation, however, was the development of a prototype system that produced an "intelligent" invoice that allowed a paper invoice to be automatically "posted."

7.4 YOU GET WHAT YOU GET

A problem that people doing ethnography often face is that there is always something more to go after. For instance, while doing ethnography, it can be very frustrating if, say, you are away for a day and come back to the organization and hear of something having occurred that you would have liked to witness. Another case is when you have done your preparatory work and decided with whom you would like to work, thought about the time to take, and then in the course of starting to work with people, you find that there are a whole number of people and work activities that you had not realized existed and that you now believe are important, but you now do not have the time to deal with. Also, you have worked extensively with someone and have moved on to work with someone else and then you see the person whose work you believed you understood, in a context you had never seen them in, doing not only something that you had not seen them doing but also something they had not told you they did. The point above was that there is always something going on, and, oftentimes, there is too much going on.

For all practical purposes, you can assume that you will not definitively cover the work site you are concerned with. However, this is neither a failure nor a problem. You can only get what you get. What you will have will be very rich and detailed and will allow you to make observations about and understand the organization of the work and the organization that you otherwise would not have done.

7.5 A LITTLE GOES A LONG WAY

A corollary of the above point is that a very small number of observations will give you much food for thought. An example here is instructive. The first National Aeronautics and Space Administration (NASA) shuttles lost a number of the tiles that made up their heat shields upon reentering the earth's atmosphere. NASA hired a professor from the engineering school at Stanford University to produce a mathematical model that would account for the failure. The professor, however, was unable to come up with a satisfactory model. She confided to us during an informal discussion of fieldwork that during a visit to the production site she was shown the area where the tiles were prepared for fixing in place. This was done using specially formulated glue. She observed that in the course of mixing the glue and applying it to the tiles, the people doing so introduce saliva into the mix after they applied it to the tile. When she asked why they did this, they explained that the glue went off very quickly making it difficult to apply the tile. However, with saliva in the mix, the glue was workable for a longer period. Of course, this changed the chemical composition of the glue and the engineer had the answer to the problem she required.

Although she was not conducting an ethnography, being able to make this direct observation of the reality of the situation was very consequential. That one small observation allowed her to overcome the problem she was grappling with. In her case, a little went an incredibly long way.

A problem that some ethnographers face is that they become obsessed by observing. Collecting data becomes an end in itself. They over identify with the people they are working with and with the setting and consider the workplace they are working in as if it was their own workplace. Although this is an extreme, sometimes in doing ethnography, too much time is spent collecting data and not enough time in analyzing it for its consequences. Given that a little can go a long way; people undertaking work practice and workplace studies should ensure that enough time is given to analyzing materials collected, integrating them through the maxims suggested above.

* * * *

CHAPTER 8

Current Status

In the past 10 years, considerable progress has been made in recognizing the importance of the social world for considerations of systems design. In many quarters of human–computer interaction, especially computer supported collaborative work, there has indeed been a move away from the individual user to the work setting as suggested by Grudin. There has been a considerable body of work assembled of actual studies, and studies of work are being used as a resource for systems design in industrial and academic environments. Studying work as a method in HCI design, however, is still something of a novelty for those areas of HCI that are not concerned with the collaborative and interactive character of human doings and is more abundant in CSCW than in mainstream HCI.

A characteristic of the use of studies of work in HCI, however, has tended to be constant throughout this decade, which is that the division of labor between the studies and the design that was a feature of early endeavors still mainly remains. Thus, those whose full-time preoccupation is with the organization of work mainly do studies of work, whereas those from within the design sciences do design that uses these studies. It is possible to find that actual designers of a system may go into the field and may even claim to be studying work in an ethnographic tradition; however, from the point of view of the type of work examined in this book, those studies are often inadequate. They either appear to be superficial, fail to render a sense of detail, or are used as post hoc rationalizations for already-made design decisions.

There are two possible developments that may be considered for the future. First, it may be possible to train people in HCI to undertake serious and rigorous studies of work, in such a way that they can use studies of work as part of their methodological arsenal. The inclusion of this book in this series on HCI method is an instantiation of this possibility, as is the fact that just as university computer science departments hire researchers in HCI, so too have they started to hire people whose main preoccupation is with studying work. If this trend continues, it is likely to result in centers of excellence in which researchers are both doing studies and design.

The second possibility is a continuation of the current trend in which there is collaboration between different disciplines within a working division of labor that articulates a design space. In this future, it is the collaboration between specialisms on a design team that is the hallmark. Studies of work done for design purposes can be articulated within the parameters of design problems and issues. This is a trend that can be evidenced in a number of academic and industrial organizations. A research question that permeates this trend, however, is one of mutually aligning the different disciplines involved so as to affect an appropriate consummation of the relationship. A method for doing this that was reported in Bentley et al. (1992) is one that is still perhaps favored today and involves the person doing the study of work, acting as a sponge absorbing the ways of the workplace, eventually to be wrung out by the designers. Whatever the future holds for this type of collaboration, it is more likely that the alignment of work studies and design will be refined in practice and good example, as opposed to methodological or theoretical fiat.

Whatever the particular future of work studies in HCI, the following is clear: HCI has had the door onto the workplace opened for it. It has been shown that the ways of work and the workplace require and make relevant interactive systems, and they are available for the design of those systems. Closing that door is not an intellectual or a practical option.

However, a further development with HCI that might appear to close the door is that of "ubiquitous computing." Those interested in this developing area rightly note that many ubiquitous systems do not support work in the workplace, but are being developed for the home, the street, on the move, and to support family life, and leisure, for example. Thus, although ethnography may still be of relevance for understanding these new settings studies of work, and the workplace would obviously not be.

Such a conclusion is, however, not correct. As we have seen in our consideration of ethnomethodology, Garfinkel points out that human action and interaction does not just happen, or is the result of forces outside of the situation, but is made to happen within the situations and circumstances in which it is done. In other words, the people doing it, as and where they do it, achieve human action, whatever it is and wherever it is done. Achieving it means that people are working to bring off their actions and interactions. Human action is worked at by those involved; it is this work that is "the missing what" of actives described in sociological literature pointed out by Garfinkel. Within ethnomethodological studies of work, "work" has always been a pun. Certainly, there has been a greater emphasis within ethnomethodology of actions and interactions that occur in the workplace, but that does not mean that studies in the genre of studies of work are not possible for actions and interactions that are not related to a job of work or the workplace. Indeed, most of the early studies in ethnomethodology and conversation analysis were not related to a workplace or job of work.

In this respect, work practice studies within HCI have a much larger remit than just the workplace, although it is the workplace and the interactions and actions that make up a job of work that has been the focus of this book. Contexting ethnomethodological studies of action and interaction for other contexts to those of the workplace remains a task for future examination.

• • • •

References

Atkinson, P., Coffey, A., Delamont, S. Lofland, J., & Lofland, L. eds. (2007). *Handbook of Ethnography*, London, Sage.

Becker, H. (1993). *Outsiders*. New York: Free Press.

Becker, H., Geer, B., Hughes, E. C., & Strauss, A. (1977). *Boys in white*. New Brunswick: Transaction Publishers.

Bentley, R., Hughes, J. A., Randall, D., Rodden, T., Sawyer, P., Shapiro, D., & Sommerville, I. (1992). Ethnographically-informed systems design for air traffic control. In *Proceedings of ACM CSCW '92 Conference on Computer-Supported Cooperative Work* (pp. 123–129).

Berger, P. L., & Luckman, T. (1966). *The social construction of reality: A treatise in the sociology of knowledge*. University of Michigan.

Blumer, H. (1969), Symbolic Interactionism: Perspective and method. Berkley: University of California Press.

Bowers, J., Button, G., & Sharrock, W. W. (1995). Workflow from within and without. *Proceedings of the ECSCW '95 Fourth European Conference on Computer-Supported Cooperative Work* (pp. 51–66). Dordrecht: Kluwer.

Button, G., Bentley, R., & Pycock, J. (2001). Lightweight computing, Palo Alto Research Centre forum.

Button, G., & Sharrock, W. W. (1997). The production of order and the order of production: Possibilities for distributed organisations, work, and technology in the print industry. *Proceedings of the Fifth European Conference on Computer Supported Cooperative Work* (pp. 1–16). Dordrecht: Kluwer.

Cole, S. (Ed.). (2001). *What's wrong with sociology*. London: Transaction Publishers.

Dourish, P., & Button, G. (1998). On "technomethodology": Foundational relationships between ethno methodology and system design. *Human–Computer Interaction* (Vol. 13, no. 4, pp. 395–432).

Durkheim, E. (1951). Suicide New York: Free Press.

Garfinkel, H. (1967a). *Studies in ethnomethodology*. Englewood Heights: Prentice-Hall.

Garfinkel, H. (Ed.). (1967b). *Ethnomethodological studies of work*. London: Routledge.

Gillespie, R. (1993). *Manufacturing knowledge: A history of the Hawthorne experiments*. Cambridge: Cambridge University Press.

Glaser, B., & Strauss, A. (1964). The social loss of dying patients, *American Journal of Nursing*, pp. 119–21.

Goffman, E. (1959). *The presentation of self in everyday life*. New York: Doubleday Anchor.

Goffman, E. (1961). *Asylums: Essays on the social situations of mental patients and other inmates*. New York: Doubleday Anchor.

Goffman, E. (1963). *Stigma: Notes on the management of spoiled identity*. New York: Simon & Schuster.

Goodwin, C., & Goodwin, M. (1993). Formulating planes: Seeing as a situated activity. In Y. Engestrom & D. Middleton (Eds.), *Communities of practice: Cognition and communication at work*. Cambridge: CUP.

Greenbaum, J., & Kyng, M. (Eds.). (1991). *Design at work*. Hillsdale: Erlbaum.

Grudin, J. (1990). The computer reaches out: The historical continuity of interface design. *Proceedings of the CHI '90 Conference on Human Factors in Computing Systems* (pp. 261–268). New York: ACM.

Heath, C., & Luff, P. (1991). Collaborative activity and technological design: Task coordination in London underground control rooms. In *Proceedings ECSCW '91* (pp. 65–79). Dordrecht: Kluwer.

Heath, C., & Luff, P. (1993). Disembodied conduct: Interactional asymmetries in video-mediated communication. In G. Button (Ed.), *Technology in working order: Studies of work, interaction and technology*. London: Routledge.

Hughes, E. C. (1971). *The sociological eye: Selected papers*. Chicago: Aldine-Atherton.

Lynch, M. (2001). Ethnomethodology and the logic of practice. In T. Schatzki et al. (Eds.), *The practice turn in contemporary theory* (pp. 131–148). Routledge: New York.

Malinowski, B. (1967). *Argonauts of the Western Pacific*. Routledge: London.

Morgan, G. (1998). Images of organization. Berrelt-Koehler publishers.

Park, R. E., Burgess, E. W., Mackenzie, R. D., & Janowitz, M. (1984). *The city: Suggestions for the investigation of human behaviour in the urban environment*. Chicago: University of Chicago Press.

Sacks, H. (1993a). *Lectures in conversation* (Vol. I). Oxford: Basil Blackwell.

Sacks, H. (1993b). *Lectures in conversation* (Vol. II). Oxford: Basil Blackwell.

Schank, R. C., & Ableson, R. F. (1977). *Scripts, plans, goals and understanding*. Hillsdale: Lawrence Erlbaum.

Schegloff, E. A. (1986). The routine as achievement. *Human Studies*, 9(2–3), pp. 111–151.

Schur, E. (1971). *Labelling deviant behaviour: Its sociological implications*. New York: Harper & Row.

Schutz, A. (1967). *Phenomenology of the social world.* Evanston, IL: Northwestern University Press.

Searle, J. (1983). *Intentionality.* Cambridge: CUP.

Sellen, A. J., & Harper, R. (2002). *The myth of the paperless office.* Net Library.

Shapiro, D. (1994). The limits of ethnography: Combining social sciences for CSCW. *Proceedings of the CSCW '94 Conference on Computer-Supported Cooperative Work* (pp. 417–428). New York: ACM.

Sharrock, W. W., & Button, G. (1997). On the relevance of Habermas' theory of communicative action for CSCW. *Computer Supported Cooperative Work,* 6(3), pp. 369–389.

Sommerville, I., Rodden, T., Sawyer, P., & Bentley, R. (1992). Sociologists can be surprisingly useful in interactive systems design. In A. Monk, D. Diaper, & M. Harrison (Eds.), *People and computers* (Vol. VII). *Proceedings of HCI '92,* pp. 341–352. Cambridge: CUP.

Suchman, L. (1987). *Plans and situated action: The problem of human–machine communication.* Cambridge: CUP.

Suchman, L. (1994). Do categories have politics? The language/action perspective reconsidered. *Computer-Supported Cooperative Work,* 2, pp. 177–190.

Thomas, W. I. (1966). *On social organisation and social personality.* Selected papers, M. Janowitz (Ed.). Chicago, IL: University of Chicago Press.

Thomas, W. I., & Znaniecki, F. (1920). *The Polish peasant in Europe and America: A monograph of an immigrant group.* University of Michigan Press.

Travers, M. (1947). The Reality of Law: Work and talk in a firm of criminal lawyers. Aldershot: Dartmouth.

Weber, M. (1964). In T. Parsons (Ed.), *The theory of social and economic organisation.* New York: Free Press.

Winograd, T., & Flores, F. (1986). *Understanding computers and cognition: A new foundation for design.* Norwood: Ablex.

Author Biographies

Graham Button gained his PhD in 1976 from the University of Manchester where he was the Faculty Research Assistant. He joined the then Plymouth Polytechnic, subsequently The University of Plymouth, in 1975 where he worked as lecturer, senior lecturer, and principal lecturer until 1992. During 1980 and 1985 he was visiting faculty at the University of California, Los Angeles and Boston University, respectively. In 1992 he joined the Cambridge laboratory of Xerox's Palo Alto Research Centre, as Principal Scientist and was appointed Director in 1999, and subsequently Laboratory Director of Xerox's European Research Centre in Grenoble, France, in 2003. In 2005 he took up the position of Executive Dean of Faculty at Sheffield Hallam University, and is now Pro-Vice Chancellor for Arts, Computing, Engineering and Sciences.

Wes Sharrock has been at the University of Manchester UK since 1965. He was a graduate student from 1965–7 then worked as assistant lecturer, lecturer, senior lecturer, reader and professor in the Department of Sociology. During 1972–3 he was a visiting associate professor at the University of British Columbia, in 1989–90 he was Visiting Senior Scientist as the Cambridge laboratory of Xerox's Palo Alto Research Centre, and in 2008 was a visiting senior scientist at the Microsoft laboratory in Cambridge UK. Current research includes studies of development work in online ontology building and of data sharing in collaborations between reaearch scientists and visualization specialists.